AF676757

Nanocomposites for Environmental and Energy Applications

Nanocomposites for Environmental and Energy Applications

Special Issue Editors

Ahmad Fauzi Ismail

Pei Sean Goh

MDPI • Basel • Beijing • Wuhan • Barcelona • Belgrade

Special Issue Editors

Ahmad Fauzi Ismail
Advanced Membrane
Technology Research Centre
School of Chemical and
Energy Engineering
Faculty of Engineering
Universiti Teknologi Malaysia
Malaysia

Pei Sean Goh
Advanced Membrane
Technology Research Centre
School of Chemical and
Energy Engineering
Faculty of Engineering
Universiti Teknologi Malaysia
Malaysia

Editorial Office
MDPI
St. Alban-Anlage 66
4052 Basel, Switzerland

This is a reprint of articles from the Special Issue published online in the open access journal *Nanomaterials* (ISSN 2079-4991) from 2018 to 2019 (available at: https://www.mdpi.com/journal/nanomaterials/special_issues/environment_energy_nano).

For citation purposes, cite each article independently as indicated on the article page online and as indicated below:

LastName, A.A.; LastName, B.B.; LastName, C.C. Article Title. *Journal Name* **Year**, *Article Number*, Page Range.

ISBN 978-3-03928-819-9 (Hbk)
ISBN 978-3-03928-820-5 (PDF)

© 2020 by the authors. Articles in this book are Open Access and distributed under the Creative Commons Attribution (CC BY) license, which allows users to download, copy and build upon published articles, as long as the author and publisher are properly credited, which ensures maximum dissemination and a wider impact of our publications.
The book as a whole is distributed by MDPI under the terms and conditions of the Creative Commons license CC BY-NC-ND.

Contents

About the Special Issue Editors

Ahmad Fauzi Ismail is a Professor at the School of Chemical and Energy Engineering, Faculty of Engineering UTM. He is the Vice Deputy Chancellor (Research and Innovation) of UTM. His research interests include the development of polymeric, inorganic and novel mixed matrix membranes for water desalination, wastewater treatment, gas separation processes, membrane for palm oil refining, photocatalytic membranes for the removal of emerging contaminants, development of hemodialysis membranes, and polymer electrolyte membrane for fuel cell applications. He obtained his Ph.D. in Chemical and Process Engineering in 1997 from the University of Strathclyde and MSc. and BSc. from Universiti Teknologi Malaysia in 1992 and 1989, respectively. His research has been published in many high impact factor journals. He has also written many academic books in this field, which have been published by reputable international publishers. He is the author and co-author of over 550 refereed journals. He has authored 7 books, 50 book chapters, 4 edited books, been granted 6 patents, and has 14 patents pending. His h-index is 75 with a cumulative citation of over 25,900.

Pei Sean Goh is an Associate Professor in the School of Chemical and Energy Engineering, Faculty of Engineering, Universiti Teknologi Malaysia (UTM). She received her Ph.D. degree in Gas Engineering in 2012 in UTM. Pei Sean is an associate research fellow of the Advanced Membrane Research Technology Research Centre (AMTEC), UTM. She is also the Head of Nanostructured Materials Research Group in UTM. Her research interests focus on the synthesis of a wide range of nanostructured materials and their composites for membrane-based separation processes. One of the main focuses of her research is the application of carbon-based nanomaterials and polymeric nanocomposite membranes for acidic gas removal as well as desalination and wastewater treatment. Pei Sean has authored or co-authored more than 130 research papers. She has also contributed 15 book chapters and 2 edited research book. Her research publication has received about 2950 citations and her current Scopus H-index is 28.

Preface to "Nanocomposites for Environmental and Energy Applications"

The exploration of various functional nanocomposite has offered vast opportunities in addressing global-scale environmental and energy issues. Over the last decade, tremendous efforts have been made in harnessing the potentials of nanocomposite through their novel synthesis and innovative applications. This Special Issue compiles a total of 9 review and research articles that deal with state-of-the-art development in nanocomposite materials for environmental and energy applications. The applications of nanocomposites in environmental remediation such as heavy metal removal and photocatalytic reaction as well as in energy applications such as supercapacitor and batteries are some of the highlights presented in this Special Issue.

Ahmad Fauzi Ismail, Pei Sean GOH

Special Issue Editors

nanomaterials

MDPI

Article

Engineering the Dimensional Interface of $BiVO_4$-2D Reduced Graphene Oxide (RGO) Nanocomposite for Enhanced Visible Light Photocatalytic Performance

Jing Sun [1,*,†], Chunxiao Wang [1,†], Tingting Shen [1], Hongchen Song [1], Danqi Li [1], Rusong Zhao [2] and Xikui Wang [3,*]

1 School of Environmental Science and Engineering, Qilu University of Technology (Shandong Academy of Sciences), Ji'nan 250353, China; wangyiwangchunxiao@163.com (C.W.); stthunanyt@163.com (T.S.); shc01261005lx@gmail.com (H.S.); autantenporte@163.com (D.L.)

2 Key Laboratory for Applied Technology of Sophisticated Analytical Instruments of Shandong Province, Analysis and Test Center, Qilu University of Technology (Shandong Academy of Sciences), Ji'nan 250014, China; zhaors1976@126.com

3 College of Environmental Science and Engineering, Shandong Agriculture and Engineering University, Ji'nan 251100, China

* Correspondence: jingsunok@163.com (J.S.); xk_wang@qlu.edu.cn (X.W.)

† These authors contributed equally to this work.

Received: 9 May 2019; Accepted: 7 June 2019; Published: 10 June 2019

Abstract: Graphene as a two-dimensional (2D) nanoplatform is beneficial for assembling a 2D heterojunction photocatalytic system to promote electron transfer in semiconductor composites. Here a $BiVO_4$ nanosheets/reduced graphene oxide (RGO) based 2D-2D heterojunction photocatalytic system as well as 0D-2D $BiVO_4$ nanoparticles/RGO and 1D-2D $BiVO_4$ nanotubes/RGO nanocomposites are fabricated by a feasible solvothermal process. During the synthesis; the growth of $BiVO_4$ and the intimate interfacial contact between $BiVO_4$ and RGO occur simultaneously. Compared to 0D-2D and 1D-2D heterojunctions, the resulting 2D-2D $BiVO_4$ nanosheets/RGO composites yield superior chemical coupling; leading to exhibit higher photocatalytic activity toward the degradation of acetaminophen under visible light irradiation. Photoluminescence (PL) and photocurrent experiments revealed that the apparent electron transfer rate in 2D-2D $BiVO_4$ nanosheets/RGO composites is faster than that in 0D-2D $BiVO_4$ nanoparticles/RGO composites. The experimental findings presented here clearly demonstrate that the 2D-2D heterojunction interface can highlight the optoelectronic coupling between nanomaterials and promote the electron–hole separation. This study will motivate new developments in dimensionality factors on designing the heterojunction photocatalysts and promote their photodegradation photocatalytic application in environmental issues.

Keywords: $BiVO_4$; RGO; visible light; two-dimensional interface; photocatalysis

1. Introduction

Semiconducting nanocrystals with tailored shapes have attracted increasing research attention in recent years due to their many intrinsic shape-dependent properties [1,2]. As an important ternary oxide semiconductor, $BiVO_4$ has been extensively investigated due to its peculiar chemical and physical functions in many fields such as dye-treatment, oxygen production, antibiotics degradation and so on [3–5]. However, the specific surface area of $BiVO_4$ is comparatively small mainly due to the large particle size [6]. The poor adsorptive performance and the poor separation efficiency of photoinduced charge carriers in pure $BiVO_4$ significantly restricts its further photocatalytic application [7].

To improve the photocatalytic performance of $BiVO_4$ photocatalysts, many approaches have been explored, such as combining with metal oxides, doping metal ions and nano-structuring [8–10].

In particular, the photocatalyst hybrids with heterojunction systems represent an effective way to enhance the photoinduced electron and holes separation. The build-in internal electric field caused by the interface of hybrids promotes the electron flow across the heterojunction [11]. Generally speaking, the electron transfer (ET) across the heterojunction interface is a key process in controlling their photocatalytic performance [12]. The challenge of assembling the heterojunction systems lies in finding an appropriate platform favorable to electron transfer between the interfaces.

Among various materials, graphene, two dimensional forms of sp^2-hybridized carbon, has exhibited outstanding characteristics such as high mechanical strength, thermal and optical properties and high electrical conductivity [13–15]. It can offer new opportunities to serve as an ideal platform to assemble the heterojunction systems. Research works have reported that the low-dimensional heterojunctions based on graphene is proven effective for ET process. For example, graphene combined with $BiVO_4$ nanoparticles or nanotubes has been synthesized and exhibits high visible-light-driven catalytic effect [7,16]. Recently, 2D dimensional heterojunctions with superior properties have motivated considerable interest in degrading pollutants. Inspired by the process for light-charge conversion in granum of green plants, 2D-2D dimensional heterojunction with $BiVO_4$ nanosheets-graphene stacked structures was fabricated to achieve rapid charge transfer [17,18]. Graphene, as an ideal 2D platform for photocatalysts assembly, benefits the electron transfer across the interface.

Recently, many literatures about $BiVO_4$ and RGO composites have been reported. However, a thoughtful and systematic comparison in $BiVO_4$-graphene nanocomposites with different dimensional heterojunctions is still scarce. Although 2D-2D dimensional heterojunctions with $BiVO_4$/graphene exhibit superior photocatalytic performance, different preparation methods make the materials incomparable. The contribution role of different $BiVO_4$ nanomaterials to enhance the composites photocatalytic activity is still unavailable. The situation may give incomplete or exaggerate information on the contribution of 2D-2D dimensional heterojunction in improving the photocatalytic performance [19]. So far, our knowledge of the specific advantages of 2D interface on developing an effective photocatalytic system is far from satisfactory.

Here a $BiVO_4$ nanosheets/RGO based 2D-2D heterojunction photocatalytic system as well as 0D-2D $BiVO_4$ nanoparticles/RGO and 1D-2D $BiVO_4$ nanotubes/RGO nanocomposites were constructed by a feasible solvothermal method. Systematic comparison with the above nanocomposites was carried out in terms of photocatalytic activity, reactive oxygen species (ROS) generation and electron transfer rate. The results emphasize the key role of interfacial dimensionality on design or fabricate graphene-semiconductor nanocomposites and improvement of the photocatalytic activity.

2. Materials and Methods

2.1. Materials

Following reagents were used: $BiCl_3$, ethanolamine and acetaminophen (Aladdin Biochemical Technology Co., Ltd., Shanghai, China), Graphite powder and NH_4VO_3 (Bodi Chemical Co., Ltd., Tianjin, China), $Bi(NO_3)_3{\cdot}5H_2O$ and ethylene diamine tetraacetic acid disodium salt (EDTA-2Na) (Tianjin Damao Chemical Reagent Co. Inc., Tianjin, China), formic acid, isopropanol (IPA) and p-Benzoquinone (Sinopharm Chemical Reagents Co., Ltd., Shanghai, China). All chemicals were used as received without further purification. Deionized water was used throughout the experiment.

2.2. Synthesis of $BiVO_4$/RGO Composites

Graphene oxide (GO) was prepared using a modified Hummers' method published in our previous work [20]. $BiVO_4$ nanosheets/RGO composites and $BiVO_4$ nanotubes/RGO composites were synthesized by the hydrothermal method, modified from previously reports [21]. Typically, 158 mg of $BiCl_3$ powder and 59 mg of NH_4VO_3 powder were added in 50 mL deionized water and stirred for 30 min to produce a homogenous suspension. Then, certain amount of 1 M ethanolamine (0.3 mL for $BiVO_4$ nanosheets; 2.5 mL for nanotubes) was added dropwise. After that, 6.32 mL 1 g L^{-1} GO solution was gradually added into the solution and then sonicated for 30 min to make the mixture uniform. The above solution was poured into a 100 mL Teflon-lined autoclave and reacted at 160 °C for 12 h. After cooling to room temperature, the resulting yellow precipitates in the reactor are collected and washed several times with alcohol and deionized water. Finally, the as prepared catalysts were dried at 60 °C for several hours. $BiVO_4$ nanosheets and $BiVO_4$ nanotubes were synthesized by a similar method without GO addition.

$BiVO_4$ nanoparticles was prepared by a modified method according to the literature [22]. The $BiVO_4$ nanoparticles/RGO composites were prepared by a hydrothermal method. The details are described in the Supplementary Materials.

2.3. Characterization

The powder X-ray diffraction (XRD) analysis was measured by Bruker-axs D-8 advance diffractometer (Cheshire, UK) with Cu Kα radiation. The morphology and element composition were recorded by using Scanning electron microscope (FE-SEM, Hitachi Regulus 8220, Tokyo, Japan). Raman measurements were acquired on a Bruker Senterra R200-L Raman spectrometer (Ettlingen, Germany). The optical adsorption behavior of the samples was performed on a Cary 5000 UV-vis-NIR spectrophotometer (Agilient Technologies, SantaClara, CA, USA). The absorption spectra were obtained by analyzing the reflectance measurement with Kubelka-Munk (KM) emission function, $F(R\infty)$. Optical band gap energy (Eg) can be determined from the plot between $E = 1240/\lambda_{Absorp.Edge}$ and $[F(R\infty)h\upsilon]^{1/2}$ where E is the photonic energy in eV and $h\upsilon$ is the energy of an incident photon. X-ray photoelectron spectroscopy (XPS) analysis was analyzed by a Kratos Axis Ultra DLD (Manchester, UK) with Al Kα X-ray source (1486.6 eV). A Fluorescence Spectrophotometer (JASCO FP-6500, Tokyo, Japan) was used for photoluminescence (PL) measurement at the excitation wavelength of 420 nm.

2.4. Photocatalytic Experiment

Acetaminophen were used as the target degradation contaminants to evaluate the photocataytic activity of the prepared catalysts. Before illumination, the solution containing 150 mL of 10 mg L^{-1} acetaminophen and 0.15 g photocatalysts was sonicated and stirred for 30 min in dark to ensure an adsorption/desorption equilibrium. Then the above suspension was irradiated by 300 W Xe arc lamp (PLS-SXE 300, Perfectlight Co. Ltd, Beijing, China) with a UV-cutoff filter ($\lambda > 400$ nm). At a given time interval of irradiation, 2 mL aliquots were collected from the suspension and centrifuged. The residual concentration of organics in the aliquots was measured by a TU-1901 UV-vis spectrophotometer (Purkinje General Instrument Co., Ltd., Beijing, China). The concentration of acetaminophen was determined by high performance liquid chromatography (LC-20AT, Shimadzu, Kyoto, Japan) with an Agela Venusil MP C18 (0.46 μm × 250 mm) reverse-phase column equipped with UV-Vis detector (SPD-20A, Shimadzu, Kyoto, Japan) at 254 nm. The mobile phase was methanol: water (35:65, v/v) and the flow rate was 0.8 mL min^{-1}. All the photocatalytic experiments were performed in triplicates and the mean values are reported.

3. Results and Discussion

3.1. Characterization of the $BiVO_4$/RGO Composites

$BiVO_4$/RGO composites were synthesized employing covalent chemistry to achieve $BiVO_4$ samples in situ growth on graphene surface. SEM and AFM images were taken to characterize the microscopic morphology and structure. Figure 1c and Figure S1 display the prepared $BiVO_4$ nanosheet exhibiting two-dimensional sheet-like morphology with 500–1000 nm in width and 8–30 nm in thickness. The products layer on the RGO sheet platform, which forms the homogenous 2D-2D interfacial contact. Figure 1b indicated that the as-prepared $BiVO_4$ nanotubes had a typical nanotubular structure and paved well on the RGO sheet to form 1D-2D heterostructures. During the synthesis, the formation of the 1D and 2D $BiVO_4$ samples was controlled by the pH value of the solution. It is reported that an increase of the pH value will decelerate the nucleation kinetics and provide a more suitable condition for anisotropic growth of 2D or 1D m-$BiVO_4$ nanostructures [21]. Similarly, the SEM images of $BiVO_4$ nanoparticles/RGO (0D-2D) exhibit the well-dispersed $BiVO_4$ nanoparticles on the RGO sheet in Figure 1a.

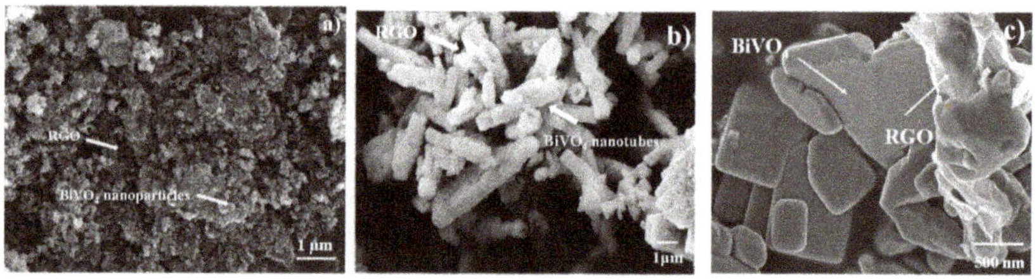

Figure 1. SEM images of $BiVO_4$/ reduced graphene oxide (RGO): (**a**) $BiVO_4$ nanoparticles/RGO; (**b**) $BiVO_4$ nanotubes/RGO; (**c**) $BiVO_4$ nanosheets/RGO.

The RGO content of the above three composites are determined at ~2 wt% by TGA test in Figure S2. According to the literature, the amount of reduced graphene oxide was determined by the weight loss from 200 to 600 °C [23].

The phase and crystal structure of the as-prepared samples were examined by XRD. As shown in Figure 2a, the XRD patterns of $BiVO_4$ and $BiVO_4$/RGO samples all agree with the monoclinic scheelite type $BiVO_4$ (JCPDS No.14-0688). Compared with $BiVO_4$ nanoparticles and nanotubes samples, the dominant 004 diffraction peak suggests that $BiVO_4$ nanosheets and $BiVO_4$ nanosheets/RGO have a preferred orientation along the (001) planes [24]. Notably, for the sample GO of Figure S3, the peak at 2θ of 10.3° is attributed to the (002) reflection of stacked GO sheets. However, no diffraction peak of GO is observed in the composites, attributing to the disappearance of layer-stacking regularity after redox of graphite [8].

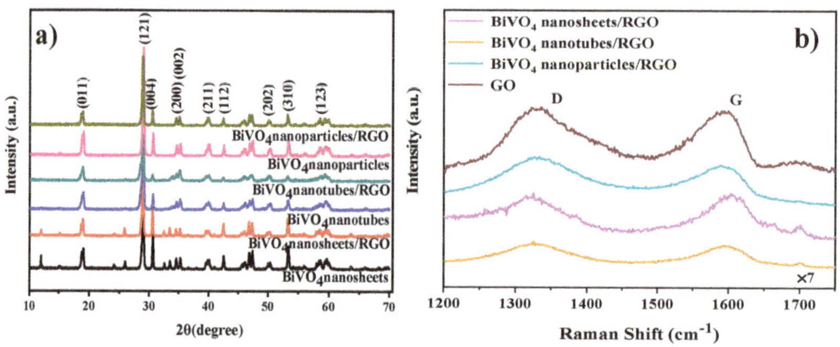

Figure 2. (**a**) XRD patterns $BiVO_4$ and $BiVO_4$/ RGO composites. (**b**) Raman spectra of graphene oxide (GO) and $BiVO_4$/RGO composites in the 1200–1800 cm^{-1} region.

The monoclinic scheelite phase of $BiVO_4$ in the composites is further confirmed by a typical Raman band at 126, 210, 325, 367, 710 and 828 cm^{-1} in Figure S4 [25]. In addition, the D band centered at 1350 cm^{-1} (disorder band) and the G band at 1580 cm^{-1} (tangential vibration band) are present, indicating that RGO has been successfully loaded on the surface of $BiVO_4$ [26,27]. Furthermore, the I_D/I_G ratio is inversely proportional to the average size of the sp^2-hybridized graphene domains [28]. As is shown in Figure 2b, after the solvothermal process, the I_D/I_G ratio of $BiVO_4$ nanosheets/RGO decreased from 1.04 to 0.99, indicating that the reduction of GO increased the average size of the graphene domains and reduced the defect density in the composite [29]. Differently, an increase in the

I_D/I_G ratio of $BiVO_4$ nanoparticles/RGO and $BiVO_4$ nanotubes/RGO are observed. The result shows that more numerous sp^2 domains have been formed in the composites [28,30].

The chemical state of $BiVO_4$ nanosheets and RGO in the composites is investigated by XPS in Figure 3. The survey spectrum in Figure 3a indicated the existence of Bi, V, O and C elements in $BiVO_4$ nanosheets/RGO. In the C1s spectrum of $BiVO_4$ nanosheets/RGO (Figure 3b), the peak centered at 283.6 eV binding energy indicates the existence of C–C bond from graphene. The peak located at 284.2 eV is attributed to the formation of C–O bond, suggesting the combination of $BiVO_4$ nanosheets and RGO. The weak O=C–O bond centered at 287.1 eV indicates that the oxygenated functional groups of reduced graphene oxide are weakened during the hydrothermal reaction, along with the reduction of GO to RGO [31]. In Figure 3c, the two peaks at 158.87 and 164.47 eV are attributed to the orbital of Bi $4f_{7/2}$ and Bi $4f_{5/2}$, indicating the existence of Bi^{3+} in the pure $BiVO_4$ nanosheets. However, in the $BiVO_4$ nanosheets/RGO composites, the binding energy of Bi $4f_{7/2}$ is blue shifted to lower values by 0.5 eV, owing to the change of the chemical environment surrounding Bi element under the influence of RGO. The same results occurred in V 2p in Figure 3d. The observations in the XPS spectra further demonstrates the intense interaction between $BiVO_4$ nanosheets and RGO.

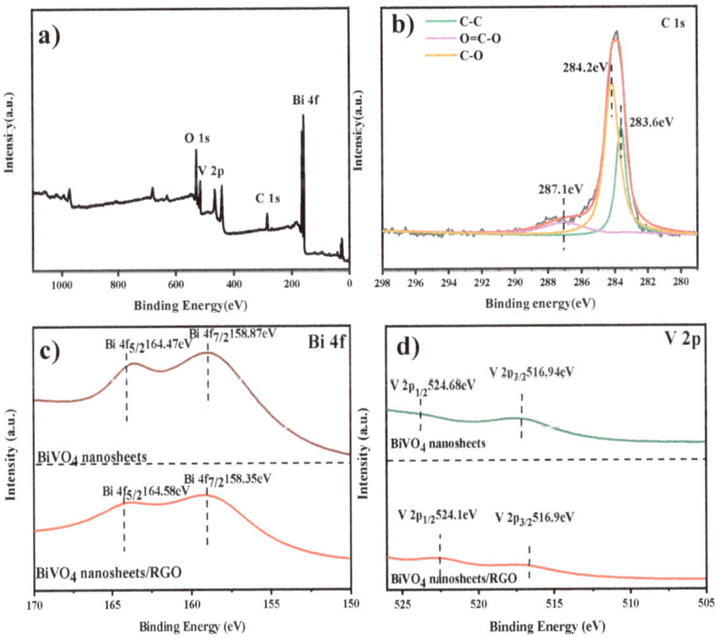

Figure 3. (**a**) The XPS survey spectrum of $BiVO_4$ nanosheets/RGO and (**b**) C 1s band of $BiVO_4$ nanosheets/RGO, (**c**) Bi 4f band of $BiVO_4$ nanosheets and $BiVO_4$ nanosheets/RGO. (**d**) V 2p band of $BiVO_4$ nanosheets and $BiVO_4$ nanosheets/RGO.

The optical properties of the composites were analyzed by DRS. The UV-vis diffuse reflectance spectra can be used to determine the absorption edge information and the width of the forbidden band of catalysts. As shown in Figure 4a, the introduction of graphene enhanced the absorbance in the visible light region for the as-prepared $BiVO_4$/RGO, especially for $BiVO_4$ nanosheets/RGO. The optical band gap energy (Eg) can be determined from the plot between $E = 1240/\lambda_{Absorp.Edge}$ and $[F(R\infty)hv]^{1/2}$ [32]. As shown in Figure 4b, the band gap energy of $BiVO_4$ nanosheets/RGO is 2.45 eV, which is lower than that of pure $BiVO_4$ nanosheets (0.21eV). These results demonstrate that the introduction of graphene in $BiVO_4$ nanosheets/RGO nanocomposites can directly produce more excited charge transfer under visible-light irradiation, which is the premise of excellent photocatalytic performance.

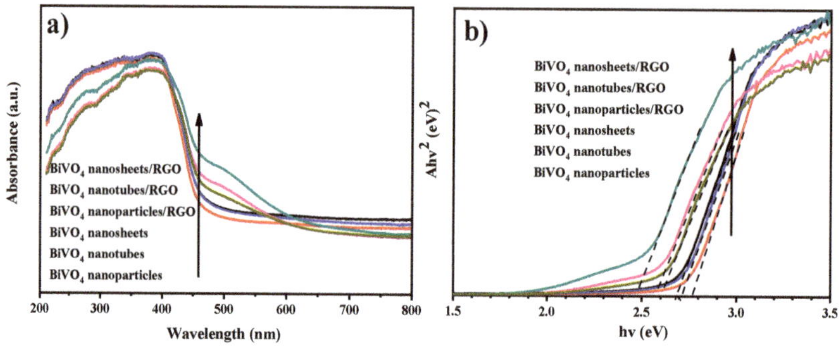

Figure 4. (**a**) UV-vis diffuse reflectance spectra of $BiVO_4$ and $BiVO_4$/RGO composites. (**b**) The relative band gap energy of the prepared samples.

3.2. Photocatalytic Performance of $BiVO_4$/RGO Samples

As a common analgesic and antipyretic drug, acetaminophen is heavily used all over the world and detected in surface water, ground water and sewage effluents [33,34]. Once acetaminophen has overdosed, it may cause potential liver damage and even death [35,36]. Thus, it is particularly urgent to provide an efficient method to enhance the degradation of acetaminophen in wastewater.

Hence, acetaminophen was used as a target pollutant to evaluate the photocatalytic properties of the prepared materials under visible light irradiation ($\lambda > 400$ nm). The chromatogram corresponding to the acetaminophen standard sample is shown in Figure S5 with the retention time at about 5.6 min. For comparison, $BiVO_4$ nanoparticles (0D), $BiVO_4$ nanotubes (1D), $BiVO_4$ nanosheets (2D), $BiVO_4$ nanoparticles/RGO (0D/2D), $BiVO_4$ nanotubes/RGO (1D/2D), and $BiVO_4$ nanosheets/RGO (2D/2D) were all examined. Figure S6 displayed the adsorption of acetaminophen on the nanomaterials reached an equilibrium state within 30 min in the dark. In Figure 5a, the photolysis performance of acetaminophen under visible light irradiation without any photocatalyst indicates that the self-degradation of acetaminophen is negligible under the visible light irradiation. The photodegradation curves of acetaminophen were fitted by pseudo-first-order reaction kinetics. As is shown in Figure 5a and Table S1, the addition of RGO can obviously improve the visible light performance of $BiVO_4$ photocatalysts, indicating the heterojunction structure between $BiVO_4$ and RGO contributed remarkably to the photocatalytic degradation rate. In particular, the $BiVO_4$ nanosheets/RGO composites ($k = 0.0141$ min^{-1}) exhibited the optimal performance compared with the corresponding pure $BiVO_4$ ($k = 0.0080$ min^{-1}). This demonstrates that the 2D-2D heterojunction structure is more beneficial to the photocatalytic activity than the other dimensional heterojunctions. When the 2D flake $BiVO_4$ and the thin slice of RGO are parallel to the space, it is beneficial to maintain high photoelectron transport efficiency and reduce the recombination of the electron hole, improving the catalytic efficiency [37].

The amphoteric behaviour of the solution influences the surface charge of the photocatalyst [38]. The role of pH on the photodegradation efficiency was studied in the pH range 3–11. As is shown in Figure S7, the photodegradation efficiency of $BiVO_4$/RGO samples increases with the increasing of pH and the maximum rate was at pH 11. That may be ascribed to major contribution of electrostatic interaction on mass transfer rate [39]. It is considered that under alkaline conditions, there is a large quantity of OH^- in the solution, which favours the formation of •OH. The strong oxidation of •OH plays an important role in the process of photocatalytic degradation [40]. Compared with $BiVO_4$ nanotubes/RGO and $BiVO_4$ nanoparticles/RGO, $BiVO_4$ nanosheets/RGO showed more excellent catalytic performance under neutral conditions, indicating that the pH application range of flaky $BiVO_4$/RGO was wider.

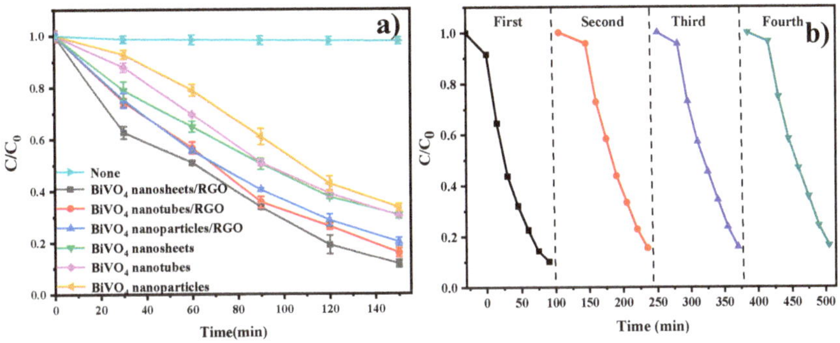

Figure 5. (**a**) Time-online photocatalytic performance of acetaminophen over the as prepared photocatalysts under visible light irradiation. (**b**) Stability experiments of $BiVO_4$ nanosheets/RGO.

The stability of the $BiVO_4$ nanosheets/RGO was evaluated by performing the recycling experiments. In Figure 5b, the photocatalysts can still maintain excellent degradation efficiency after four cycles. Figure 6a displays the smooth surface of $BiVO_4$ nanosheet after visible light irradiation. Compared with the fresh sample, no obvious discrepancy in the XRD pattern of the recycled sample was observed in Figure 6b. As is shown in Figures 3b and 6c, the XPS spectra of the recycled $BiVO_4$ nanosheets/RGO exhibited a slight decrease of the C–O and O=C–O peak intensity, possibly attributed to the further photocatalytic reduction of GO to RGO during the photocatalytic degradation [20]. However, the above changes did not affect the photocatalytic performance after recycling experiments. It is proved that the $BiVO_4$ nanosheets/RGO composites prepared exhibit relatively high stability.

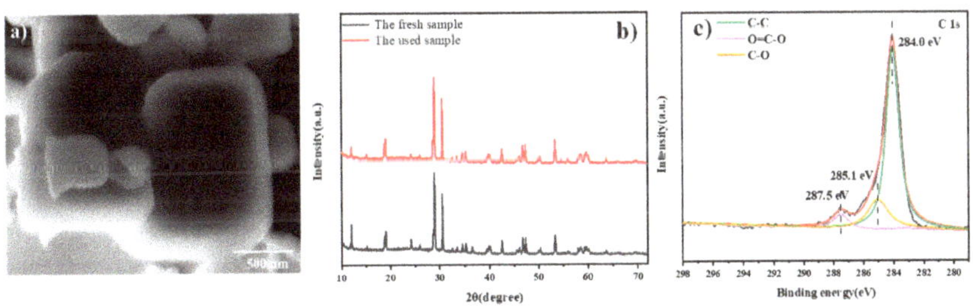

Figure 6. (**a**) SEM, (**b**) XRD and (**c**) XPS results of C 1s band of $BiVO_4$ nanosheets/RGO after four cycles irradiation.

3.3. *Photocatalytic Mechanism of BiVO$_4$/RGO*

It is commonly accepted that a series of reactive species, such as hole (h^+), hydroxyl radical (•OH), and superoxide radial ($O_2^{\bullet-}$), usually govern the photocatalytic degradation reactions of organic pollutants [41]. In order to investigate the main reactive species responsible for the photocatalytic activity of $BiVO_4$/RGO samples, the radical trapping experiment was carried out. Figure 7 displays the photocatalytic degradation curves of acetaminophen over $BiVO_4$ nanotubes/RGO, $BiVO_4$ nanoparticles/RGO, and $BiVO_4$ nanosheets/RGO with the addition of ROS scavengers. In this experiment, isopropanol (IPA) was used to quench •OH, formic acid for h^+, and p-benzoquinone (BQ) for $O_2^{\bullet-}$. For comparison, the radical trapping results of the pure $BiVO_4$ samples was demonstrated in Figure S8.

As depicted in Figure 7b,c, acetaminophen degradation process was obviously depressed by IPA, verifying •OH plays the most important role in the photocatalytic reaction over $BiVO_4$ nanosheets/RGO and $BiVO_4$ nanotubes/RGO. In addition, when the scavenger for h^+ was added into the photocatalytic solution in Figure 7c, the degradation of acetaminophen was also depressed. It illustrates that

h^+ was involved as minor radical species in the photocatalytic process of $BiVO_4$ nanosheets/RGO. As for the system based on $BiVO_4$ nanoparticles/RGO, the reaction process was a little different. The degradation rate of acetaminophen in Figure 7a showed a certain decrease in the presence of formic acid, indicating h^+ is the key reactive species responsible for the photodegradation over $BiVO_4$ nanoparticles/RGO. According to the results mentioned above, we can preliminarily conclude that for $BiVO_4$ nanoparticles/RGO photocatalytic process, h^+ is the main radical species. In the $BiVO_4$ nanotubes/RGO reaction system, •OH plays an important role, while in the $BiVO_4$ nanosheets/RGO based reaction system, •OH and h^+ as the major and minor radical species are all produced and participated in the acetaminophen degradation process.

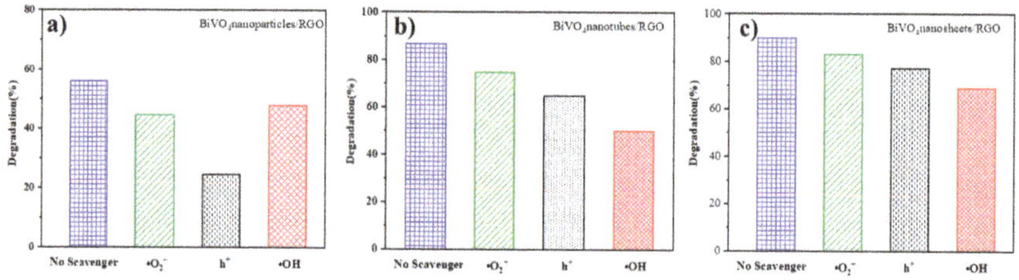

Figure 7. Free radical inhibition experiment of $BiVO_4$/RGO: (**a**) $BiVO_4$ nanoparticles/ RGO; (**b**) $BiVO_4$ nanotubes/RGO; (**c**) $BiVO_4$ nanosheets/RGO.

Different from the $BiVO_4$/RGO photocatalytic systems, pure $BiVO_4$ samples take on the similar characteristics in Figure S8. For the pure $BiVO_4$ nanoparticles, nanotubes and nanosheets, h^+ is the main radical species participating the photocatalytic processes.

The involved ROS is further confirmed by ESR experiments under visible light. As is shown in Figure 8b, after the catalysts were exposed to visible light for 10 min, the characteristic peaks of h^+ increased obviously for both $BiVO_4$ nanoparticles/RGO and $BiVO_4$ nanosheets/RGO than in dark condition, and the signals for $BiVO_4$ nanotubes/RGO could nearly be ignored. Notably, the peak intensity of h^+ referring to the $BiVO_4$ nanosheets/RGO was much higher than that of $BiVO_4$ nanoparticles/RGO. The result validated that the generation amount of h^+ of this 2D-2D system was more than other nanocomposites. Moreover, from Figure 8c, obvious signals of Hydroxy-5, 5-dimethyl-1-pyrrolidinyloxy (DMPO-•OH) were also observed for $BiVO_4$ nanotubes/RGO and $BiVO_4$ nanosheets/RGO in the measurement under visible light irradiation, implying that •OH was produced in the above two reaction systems and took part in the photocatalytic process. It is noticeable that although stronger intensity DMPO- $O_2^{\bullet-}$ and DMPO-•OH adducts were found in $BiVO_4$ nanotubes/RGO system, $BiVO_4$ nanosheets/RGO displayed higher photocatalytic efficiency for acetaminophen. This suggests that the photogenerated valence band hole of $BiVO_4$ nanosheets/RGO can oxidize water to generate •OH and participated in the acetaminophen degradation process [42].

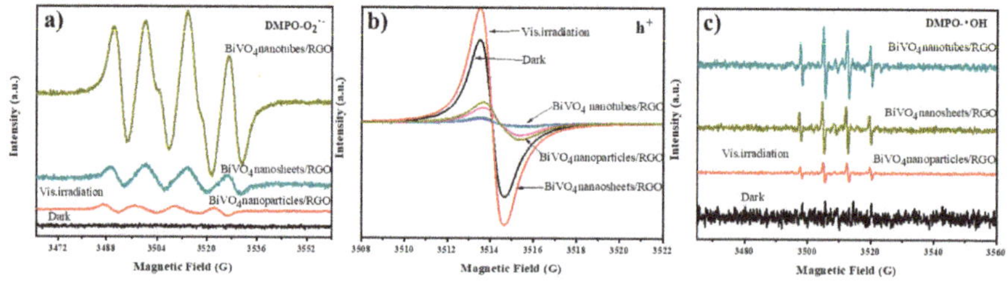

Figure 8. Electron spin resonance spectra of radical in $BiVO_4$/RGO composites under visible light: (**a**) DMPO-$O_2^{\bullet-}$, (**b**) h^+ and (**c**) DMPO-•OH.

For comparison, the electron spin resonance spectra of the pure $BiVO_4$ samples was demonstrated in Figure S9. Although the stronger signals of h^+ and DMPO-O_2•$^-$ were observed with $BiVO_4$ nanosheets, the photocatalytic degradation efficiency is similar to the $BiVO_4$ nanotubes in Figure 5a and Table S1. It can be inferred that, although the photocatalytic activity of $BiVO_4$ materials decreases dramatically after addition of formic acid, •OH dominants the acetaminophen degradation process. The inhibition of holes reduces the amount of reactive hydroxyl radicals, thereby reducing the photocatalytic activity of the system. Thus, this is to say, h^+ as well as •OH was the main active species participating in the pure $BiVO_4$ photocatalytic system.

It was reported that higher separation efficiency of electron-hole pairs plays a vital role in the photocatalytic degradation of pollutants [43,44]. According to the radical trapping experiment and ESR analysis, the reaction mechanism of 1D and 2D $BiVO_4$/RGO heterojunctions for degrading organic pollutants was proposed (Scheme 1). It is assumed that 2D-2D heterojunction between $BiVO_4$ nanosheets and graphene can facilitate the photogenerated electron transfer. That may promote the direct participation of holes in the photocatalytic process or the reaction with OH^- to generate •OH. 1D-2D heterojunction interfaces have the ability to yield more photo-generated electrons in the degradation process and promotes the oxygen molecules to generate $O_2^{\bullet -}$ and then oxidized to get •OH. Besides, the stronger intensity of h^+, $O_2^{\bullet -}$, and •OH in 2D-2D system also demonstrates the intense interface facilitates more efficient charge separation and transfer.

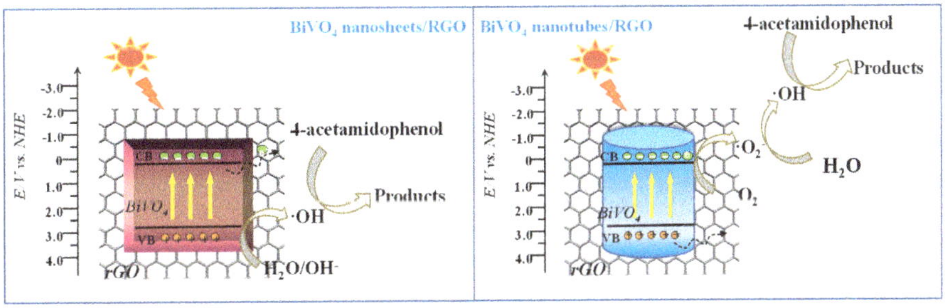

Scheme 1. Schematic image of electron-hole separation mechanism for $BiVO_4$/RGO.

3.4. Photoinduced Electron Transfer Properties of $BiVO_4$/RGO Composites

In a photo-degradation process, the higher photocatalytic efficiency demands that the electron transfer is faster than the recombination [45]. The prolonged lifetime of the photogenerated electrons can be supported by Photoluminescence (PL) spectra in Figure S10. Under an excitation wavelength of 420 nm, the main emission peak of $BiVO_4$ was detected at around 521 nm, owing to the recombination of electrons in the conduction band and holes in the valence band [46]. The introduction of graphene quenched the PL intensity of excited $BiVO_4$ nanocomposites. The orders of the detected PL intensities were: $BiVO_4$ nanoparticles > $BiVO_4$ nanotubes > $BiVO_4$ nanosheets > $BiVO_4$ nanoparticles/RGO > $BiVO_4$ nanotubes/RGO > $BiVO_4$ nanosheets/RGO, which was in good accordance with the photocatalytic behaviors. Furthermore, the lower PL intensity of $BiVO_4$ nanosheets/RGO suggests that the recombination of the photogenerated electron-hole pairs is efficiently inhibited by the two-dimensional heterjunction interface and the charge carriers separation rate is promoted.

The enhanced charge transfer rate of $BiVO_4$ nanosheets/RGO was further demonstrated by the transient photocurrent responses. As displayed in Figure 9, the photocurrent density of the $BiVO_4$/RGO composites is much higher than that of the pure $BiVO_4$ samples, especially for $BiVO_4$ nanosheets/RGO. That is in good accordance with the result of photocatalytic performances. It is indicated that the enhanced photocurrent response of the $BiVO_4$ nanosheets/RGO represents higher efficiency of charge separation and lower recombination rate in 2D-2D heterojunction interface [47].

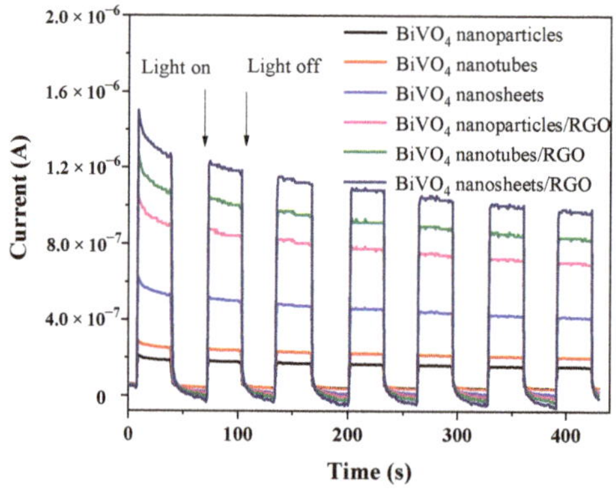

Figure 9. Photocurrent responses of the prepared nanocomposites in 0.5 M Na_2SO_4 solution during the repeated on-off cycles under visible light irradiation.

On the basis of the above results, the charge transfer mechanism could be proposed as follows. Due to the intimate contact of 2D-2D interface, the favorable transfer of electrons from $BiVO_4$ nanosheets to graphene can reduce the recombination of electron–hole pairs. The enhanced generation of ROS, especially h^+ and •OH, accelerate the photocatalytic degradation process of acetaminophen.

4. Conclusions

In summary, the $BiVO_4$ nanoparticles/RGO, $BiVO_4$ nanotubes/RGO and $BiVO_4$ nanosheets/RGO hybrids were prepared and the photocatalytic performance was evaluated. The morphology, chemical structures and photocatalytic performance of the as-prepared samples are studied through a series of characterization. Compared to 0D/2D and 1D/2D nanocomposites, 2D/2D $BiVO_4$ nanosheets/RGO with two-dimensional interface exhibits higher photoactivity. That can be attributed to a stronger electronic and physical coupling effect between $BiVO_4$ nanosheets and graphene nanosheets, which allows for the prolonged lifetime and effective separation of electrons and holes. The 2D-2D heterojunction interface opens a new window for exploiting a visible light photocatalytic system with well-defined nanohybrids to purify the polluted environment.

Supplementary Materials: The following are available online at http://www.mdpi.com/2079-4991/9/6/907/s1, the details of preparation of $BiVO_4$ nanoparticles and $BiVO_4$ nanoparticles/RGO, Figure S1: Atomic force microcopy images of the 2D $BiVO_4$ nanosheets, Figure S2: Thermo gravimetric analysis (TGA) of the $BiVO_4$/RGO composites, Figure S3: XRD patterns of the prepared graphene oxide (GO), Figure S4: Raman spectra of $BiVO_4$ and $BiVO_4$/RGO composites, Figure S5: HPLC chromatograms of acetaminophen standard sample, Figure S6: The adsorptive performance of acetaminophen over the $BiVO_4$ and $BiVO_4$/RGO composites without visible light irradiation, Table S1. The pseudo-first order kinetic equation and the rate constant (k) of BiVO4 and BiVO4/RGO composites, Figure S7: Photocatalytic degradation of RhB over photocatalysts under different pH conditions: (a) $BiVO_4$ nanosheet/RGO; (b) $BiVO_4$ nanotube/RGO;(c) $BiVO_4$ nanoparticle/RGO, Figure S8: Free radical inhibition experiment of $BiVO_4$ samples: (a) $BiVO_4$ nanoparticles; (b) $BiVO_4$ nanotubes; (c) $BiVO_4$ nanosheets, Figure S9: Electron spin resonance spectra of radical in $BiVO_4$ samples under visible light: (a) DMPO-O_2•$^-$, (b) h^+ and (c) DMPO-•OH, Figure S10: PL spectra of $BiVO_4$ and $BiVO_4$/RGO samples.

Author Contributions: Conceptualization, J.S. and X.W.; performed the experiments, C.W.; analyzed the data, J.S., H.S. and D.L.; writing-original draft preparation, J.S. and X.W.; writing—review and editing, T.S. and X.W.; funding acquisition, R.Z. All authors read and approved the final version of the manuscript.

Funding: This research was funded by the National Nature Science Foundation of China (No. 21507067), Doctoral Found for Cooperation Projects of Qilu University of Technology (Shandong Academy of Sciences) (No.2017BSHZ019) and International Cooperation Research Special Funds Project of Qilu University of Technology (Shandong Academy of Sciences) (No. QLUTGJHZ2018004).

Conflicts of Interest: The authors declare no conflict of interest.

References

1. Yang, H.G.; Liu, G.; Qiao, S.Z.; Sun, C.H.; Jin, Y.G.; Smith, S.C.; Zou, J.; Cheng, H.M.; Lu, G.Q. Solvothermal Synthesis and Photoreactivity of Anatase TiO_2 Nanosheets with Dominant {001} Facets. *J. Am. Chem. Soc.* **2009**, *131*, 4078–4083. [CrossRef] [PubMed]
2. Jun, Y.-w.; Choi, J.-S.; Cheon, J. Shape Control of Semiconductor and Metal Oxide Nanocrystals through Nonhydrolytic Colloidal Routes. *Angew. Chem. Int. Ed.* **2006**, *45*, 3414–3439. [CrossRef] [PubMed]
3. Kudo, A.; Omori, K.; Kato, H. A novel aqueous process for preparation of crystal form-controlled and highly crystalline $BiVO_4$ powder from layered vanadates at room temperature and its photocatalytic and photophysical properties. *J. Am. Chem. Soc.* **1999**, *121*, 11459–11467. [CrossRef]
4. Zhao, Y.; Xie, Y.; Zhu, X.; Yan, S.; Wang, S. Surfactant-free synthesis of hyperbranched monoclinic bismuth vanadate and its applications in photocatalysis, gas sensing, and lithium-ion batteries. *Chemistry* **2008**, *14*, 1601–1606. [CrossRef] [PubMed]
5. Antuch, M.; Millet, P.; Iwase, A.; Kudo, A. The role of surface states during photocurrent switching: Intensity modulated photocurrent spectroscopy analysis of $BiVO_4$ photoelectrodes. *Appl. Catal. B Environ.* **2018**, *237*, 401–408. [CrossRef]
6. Bian, J.; Qu, Y.; Zhang, X.L.; Sun, N.; Tang, D.Y.; Jing, L.Q. Dimension-matched plasmonic Au/TiO_2/$BiVO_4$ nanocomposites as efficient wide-visible-light photocatalysts to convert CO_2 and mechanistic insights. *J. Mater. Chem. A* **2018**, *6*, 11838–11845. [CrossRef]
7. Yan, Y.; Sun, S.F.; Song, Y.; Yan, X.; Guan, W.S.; Liu, X.L.; Shi, W.D. Microwave-assisted in situ synthesis of reduced graphene oxide-$BiVO_4$ composite photocatalysts and their enhanced photocatalytic performance for the degradation of ciprofloxacin. *J. Hazard. Mater.* **2013**, *250*, 106–114. [CrossRef]
8. Yu, Q.Q.; Tang, Z.R.; Xu, Y.J. Synthesis of $BiVO_4$ nanosheets-graphene composites toward improved visible light photoactivity. *J. Energy Chem.* **2014**, *23*, 564–574. [CrossRef]
9. Jiang, H.Q.; Endo, H.; Natori, H.; Nagai, M.; Kobayashi, K. Fabrication and efficient photocatalytic degradation of methylene blue over CuO/$BiVO_4$ composite under visible-light irradiation. *Mater. Res. Bull.* **2009**, *44*, 700–706. [CrossRef]
10. Wang, F.X.; Shao, M.W.; Cheng, L.; Hua, J.; Wei, X.W. The synthesis of monoclinic bismuth vanadate nanoribbons and studies of photoconductive, photoresponse, and photocatalytic properties. *Mater. Res. Bull.* **2009**, *44*, 1687–1691. [CrossRef]
11. Yu, W.J.; Liu, Y.; Zhou, H.L.; Yin, A.X.; Li, Z.; Huang, Y.; Duan, X.F. Highly efficient gate-tunable photocurrent generation in vertical heterostructures of layered materials. *Nat. Nanotechnol.* **2013**, *8*, 952–958. [CrossRef] [PubMed]
12. Tachikawa, T.; Wang, N.; Yamashita, S.; Cui, S.C.; Majima, T. Design of a Highly Sensitive Fluorescent Probe for Interfacial Electron Transfer on a TiO_2 Surface. *Angew. Chem. Int. Ed.* **2010**, *49*, 8593–8597. [CrossRef] [PubMed]
13. Fang, Y.X.; Guo, S.J.; Zhu, C.Z.; Zhai, Y.M.; Wang, E.K. Self-Assembly of Cationic Polyelectrolyte-Functionalized Graphene Nanosheets and Gold Nanoparticles: A Two-Dimensional Heterostructure for Hydrogen Peroxide Sensing. *Langmuir* **2010**, *26*, 11277–11282. [CrossRef] [PubMed]
14. Kwak, J.Y.; Hwang, J.; Calderon, B.; Alsalman, H.; Munoz, N.; Schutter, B.; Spencer, M.G. Electrical Characteristics of Multilayer MoS_2 FET's with MoS_2/Graphene Heterojunction Contacts. *Nano Lett.* **2014**, *14*, 4511–4516. [CrossRef] [PubMed]
15. Chen, F.; Yang, Q.; Zhong, Y.; An, H.X.; Zhao, J.W.; Xie, T.; Xu, Q.X.; Li, X.M.; Wang, D.B.; Zeng, G.M. Photo-reduction of bromate in drinking water by metallic Ag and reduced graphene oxide (RGO) jointly modified $BiVO_4$ under visible light irradiation. *Water Res.* **2016**, *101*, 555–563. [CrossRef] [PubMed]
16. Sun, Y.F.; Qu, B.Y.; Liu, Q.; Gao, S.; Yan, Z.X.; Yan, W.S.; Pan, B.C.; Wei, S.Q.; Xie, Y. Highly efficient visible-light-driven photocatalytic activities in synthetic ordered monoclinic $BiVO_4$ quantum tubes-graphene nanocomposites. *Nanoscale* **2012**, *4*, 3761–3767. [CrossRef] [PubMed]
17. Li, Y.; Sun, Z.; Zhu, S.; Liao, Y.; Chen, Z.; Zhang, D. Fabrication of $BiVO_4$ nanoplates with active facets on graphene sheets for visible-light photocatalyst. *Carbon* **2015**, *94*, 599–606. [CrossRef]
18. Tang, Z.R.; Yu, Q.Q.; Xu, Y.J. Toward improving the photocatalytic activity of $BiVO_4$-graphene 2D-2D composites under visible light by the addition of mediator. *RSC Adv.* **2014**, *4*, 58448–58452. [CrossRef]

19. Zhang, Y.H.; Tang, Z.R.; Fu, X.; Xu, Y.J. Engineering the Unique 2D Mat of Graphene to Achieve Graphene-TiO_2 Nanocomposite for Photocatalytic Selective Transformation: What Advantage does Graphene Have over Its Forebear Carbon Nanotube? *ACS Nano* **2011**, *5*, 7426–7435. [CrossRef]
20. Sun, J.; Zhang, H.; Guo, L.H.; Zhao, L.X. Two-Dimensional Interface Engineering of a Titania-Graphene Nanosheet Composite for Improved Photocatalytic Activity. *ACS Appl. Mater. Interfaces* **2013**, *5*, 13035–13041. [CrossRef]
21. Cao, S.W.; Yin, Z.; Barber, J.; Boey, F.Y.; Loo, S.C.; Xue, C. Preparation of Au-$BiVO_4$ heterogeneous nanostructures as highly efficient visible-light photocatalysts. *ACS Appl. Mater. Interfaces* **2012**, *4*, 418–423. [CrossRef] [PubMed]
22. Ke, D.; Peng, T.; Ma, L.; Cai, P.; Jiang, P. Photocatalytic water splitting for O_2 production under visible-light irradiation on $BiVO_4$ nanoparticles in different sacrificial reagent solutions. *Appl. Catal. A Gen.* **2008**, *350*, 111–117. [CrossRef]
23. Yu, S.H.; Conte, D.E.; Baek, S.; Lee, D.C.; Park, S.K.; Lee, K.J.; Piao, Y.; Sung, Y.E.; Pinna, N. Structure-Properties Relationship in Iron Oxide-Reduced Graphene Oxide Nanostructures for Li-Ion Batteries. *Adv. Funct. Mater.* **2013**, *23*, 4293–4305. [CrossRef]
24. Xi, G.; Ye, J. Synthesis of bismuth vanadate nanoplates with exposed {001} facets and enhanced visible-light photocatalytic properties. *Chem. Commun.* **2010**, *46*, 1893–1895. [CrossRef] [PubMed]
25. Chen, Y.X.; Ma, X.G.; Li, D.; Wang, H.H.; Huang, C.Y. Mechanism of enhancing visible-light photocatalytic activity of $BiVO_4$ via hybridization of graphene based on a first-principles study. *RSC Adv.* **2017**, *7*, 4395–4401. [CrossRef]
26. Dong, S.Y.; Cui, Y.R.; Wang, Y.F.; Li, Y.K.; Hu, L.M.; Sun, J.Y.; Sun, J.H. Designing three-dimensional acicular sheaf shaped $BiVO_4$/reduced graphene oxide composites for efficient sunlight-driven photocatalytic degradation of dye wastewater. *Chem. Eng. J.* **2014**, *249*, 102–110. [CrossRef]
27. Fu, Y.S.; Sun, X.Q.; Wang, X. $BiVO_4$-graphene catalyst and its high photocatalytic performance under visible light irradiation. *Mater. Chem. Phys.* **2011**, *131*, 325–330. [CrossRef]
28. Wang, X.T.; Ling, D.D.; Wang, Y.M.; Long, H.; Sun, Y.B.; Shi, Y.Q.; Chen, Y.C.; Jing, Y.; Sun, Y.M.; Dai, Y.Q. N-doped graphene quantum dots-functionalized titanium dioxide nanofibers and their highly efficient photocurrent response. *J. Mater. Res.* **2014**, *29*, 1408–1416. [CrossRef]
29. Chen, S.; Zhu, J.W.; Wang, X. One-Step Synthesis of Graphene-Cobalt Hydroxide Nanocomposites and Their Electrochemical Properties. *J. Phys. Chem. C* **2010**, *114*, 11829–11834. [CrossRef]
30. Mai, J.W.; Liu, W.; Qiu, J.L.; Wu, F.J.; Liu, H.F.; Zhou, W.Y.; Fang, Y.P.; Zhang, S.T. Characterization and Enhanced Visible-Light Photocatalytic Properties of {001} Facets-Exposed TiO_2-Reduced Graphene Oxide Nanocomposites. *J. Nanosci. Nanotechnol.* **2015**, *15*, 4870–4876. [CrossRef]
31. Kwon, W.; Kim, Y.H.; Lee, C.L.; Lee, M.; Choi, H.C.; Lee, T.W.; Rhee, S.W. Electroluminescence from graphene quantum dots prepared by amidative cutting of tattered graphite. *Nano Lett.* **2014**, *14*, 1306–1311. [CrossRef] [PubMed]
32. Wetchakun, N.; Chaiwichain, S.; Inceesungvorn, B.; Pingmuang, K.; Phanichphant, S.; Minett, A.I.; Chen, J. $BiVO_4/CeO_2$ Nanocomposites with High Visible-Light-Induced Photocatalytic Activity. *ACS Appl. Mater. Interfaces* **2012**, *4*, 3718–3723. [CrossRef] [PubMed]
33. Roberts, P.H.; Thomas, K.V. The occurrence of selected pharmaceuticals in wastewater effluent and surface waters of the lower Tyne catchment. *Sci. Total Environ.* **2006**, *356*, 143–153. [CrossRef] [PubMed]
34. Zwiener, C. Occurrence and analysis of pharmaceuticals and their transformation products in drinking water treatment. *Anal. Bioanal. Chem.* **2007**, *387*, 1159–1162. [CrossRef] [PubMed]
35. Su, C.C.; Cada, C.A.; Dalida, M.L.P.; Lu, M.C. Effect of UV light on acetaminophen degradation in the electro-Fenton process. *Sep. Purif. Technol.* **2013**, *120*, 43–51. [CrossRef]
36. Rajoriya, S.; Bargole, S.; George, S.; Saharan, V.K.; Gogate, P.R.; Pandit, A.B. Synthesis and characterization of samarium and nitrogen doped TiO_2 photocatalysts for photo-degradation of 4-acetamidophenol in combination with hydrodynamic and acoustic cavitation. *Sep. Purif. Technol.* **2019**, *209*, 254–269. [CrossRef]
37. Beranek, R. (Photo)electrochemical Methods for the Determination of the Band Edge Positions of TiO_2-Based Nanomaterials. *Adv. Phys. Chem.* **2011**, *2011*, 20. [CrossRef]
38. Sakthivel, S.; Neppolian, B.; Shankar, M.V.; Arabindoo, B.; Palanichamy, M.; Murugesan, V. Solar photocatalytic degradation of azo dye: comparison of photocatalytic efficiency of ZnO and TiO_2. *Sol. Energy Mater. Sol. Cells* **2003**, *77*, 65–82. [CrossRef]

39. Mosleh, S.; Rahimi, M.R.; Ghaedi, M.; Dashtian, K. Sonophotocatalytic degradation of trypan blue and vesuvine dyes in the presence of blue light active photocatalyst of Ag_3PO_4/Bi_2S_3-HKUST-1-MOF: Central composite optimization and synergistic effect study. *Ultrason. Sonochem.* **2016**, *32*, 387–397. [CrossRef]
40. Sakthivel, S.; Neppoiian, B.; Palanichamy, M.; Arabindoo, B.; Murugesan, V. Photocatalytic degradation of leather dye over ZnO catalyst supported on alumina and glass surfaces. *Water Sci. Technol.* **2001**, *44*, 211–218. [CrossRef]
41. Zhang, S.; Su, C.; Ren, H.; Li, M.; Zhu, L.; Ge, S.; Wang, M.; Zhang, Z.; Li, L.; Cao, X. In-Situ Fabrication of g-C_3N_4/ZnO Nanocomposites for Photocatalytic Degradation of Methylene Blue: Synthesis Procedure Does Matter. *Nanomaterials* **2019**, *9*, 215. [CrossRef] [PubMed]
42. Palominos, R.; Freer, J.; Mondaca, M.A.; Mansilla, H.D. Evidence for hole participation during the photocatalytic oxidation of the antibiotic flumequine. *J. Photochem. Photobiol. A* **2008**, *193*, 139–145. [CrossRef]
43. Chen, F.; Yang, Q.; Li, X.M.; Zeng, G.M.; Wang, D.B.; Niu, C.G.; Zhao, J.W.; An, H.X.; Xie, T.; Deng, Y.C. Hierarchical assembly of graphene-bridged $Ag_3PO_4/Ag/BiVO_4$ (040) Z-scheme photocatalyst: An efficient, sustainable and heterogeneous catalyst with enhanced visible-light photoactivity towards tetracycline degradation under visible light irradiation. *Appl. Catal. B Environ.* **2017**, *200*, 330–342. [CrossRef]
44. Si, Y.H.; Xia, Y.; Shang, S.K.; Xiong, X.B.; Zeng, X.R.; Zhou, J.; Li, Y.Y. Enhanced Visible Light Driven Photocatalytic Behavior of $BiFeO_3$/Reduced Graphene Oxide Composites. *Nanomaterials* **2018**, *8*, 526. [CrossRef] [PubMed]
45. Zhang, H.; Lv, X.J.; Li, Y.M.; Wang, Y.; Li, J.H. P25-Graphene Composite as a High Performance Photocatalyst. *Acs Nano* **2010**, *4*, 380–386. [CrossRef] [PubMed]
46. Sun, H.Q.; Zhou, G.L.; Wang, Y.X.; Suvorova, A.; Wang, S.B. A New Metal-Free Carbon Hybrid for Enhanced Photocatalysis. *ACS Appl. Mater. Interfaces* **2014**, *6*, 16745–16754. [CrossRef] [PubMed]
47. Gao, X.H.; Bin Wu, H.B.; Zheng, L.X.; Zhong, Y.J.; Hu, Y.; Lou, X.W. Formation of Mesoporous Heterostructured $BiVO_4/Bi_2S_3$ Hollow Discoids with Enhanced Photoactivity. *Angew. Chem. Int. Ed.* **2014**, *53*, 5917–5921. [CrossRef] [PubMed]

© 2019 by the authors. Licensee MDPI, Basel, Switzerland. This article is an open access article distributed under the terms and conditions of the Creative Commons Attribution (CC BY) license (http://creativecommons.org/licenses/by/4.0/).

nanomaterials

MDPI

Article

Cu-Doped Porous Carbon Derived from Heavy Metal-Contaminated Sewage Sludge for High-Performance Supercapacitor Electrode Materials

Zhouliang Tan [1], Feng Yu [1], Liu Liu [1], Xin Jia [1], Yin Lv [1], Long Chen [1], Yisheng Xu [1,2], Yulin Shi [1,*] and Xuhong Guo [1,2]

1 Key Laboratory for Green Processing of Chemical Engineering of Xinjiang Bingtuan, School of Chemistry and Chemical Engineering, Shihezi University, Shihezi 832003, China; tzl6880@foxmail.com (Z.T.); yufeng05@mail.ipc.ac.cn (F.Y.); liuliu66shzu@163.com (L.L.); jiaxin_shzu@foxmail.com (X.J.); ag_125@163.com (Y.L.); chenlong2012@sinano.ac.cn (L.C.); ecustn@gmail.com (Y.X.); guoxuhong@ecust.edu.cn (X.G.)

2 State Key Laboratory of Chemical Engineering, East China University of Science and Technology, Shanghai 200237, China

* Correspondence: shiyulin521@126.com; Tel.: +86-993-205-5030

Received: 25 April 2019; Accepted: 11 June 2019; Published: 17 June 2019

Abstract: In this paper, we report a complete solution for enhanced sludge treatment involving the removal of toxic metal (Cu(II)) from waste waters, subsequent pyrolytic conversion of these sludge to Cu-doped porous carbon, and their application in energy storage systems. The morphology, composition, and pore structure of the resultant Cu-doped porous carbon could be readily modulated by varying the flocculation capacity of Cu(II). The results demonstrated that it exhibited outstanding performance for supercapacitor electrode applications. The Cu(II) removal efficiency has been evaluated and compared to the possible energy benefits. The flocculant dosage up to 200 $mg \cdot L^{-1}$ was an equilibrium point existing between environmental impact and energy, at which more than 99% Cu(II) removal efficiency was achieved, while the resulting annealed product showed a high specific capacity (389.9$\cdot F \cdot g^{-1}$ at $1 \cdot A \cdot g^{-1}$) and good cycling stability (4% loss after 2500 cycles) as an electrode material for supercapacitors.

Keywords: heavy metal; flocculation; Cu-doped carbon; supercapacitor

1. Introduction

The heavy metal contamination resulting from mining, smelting activities, exhaust gas discharge, and sewage irrigation has detrimental effects on human health and environmental sustainability. Recent studies have consistently shown an association between heavy metal pollution and physical discomfort, and even death, in humans [1]. However, because of their poor degradability, dissolved heavy metal ions normally need to be physically removed from wastewater by immobilization techniques such as ion exchange, flocculation, membrane filtration, and adsorption [2]. Even though these approaches aid the removal of heavy metals from wastewaters, they generate large quantities of sewage sludge. It is estimated that over 8×10^6 tons of sewage sludge are generated each day from wastewater treatment plants in China [3],which takes further issue in sludge handling and final disposal.

As an important residue produced from wastewater treatment, sewage sludge recovery and recycling are always considered as an adequate challenge in wastewater management [1]. Sewage sludge landfill, incineration processes, soil application, and dumping at sea are considered as the most effective methods for the disposal of waste sludge in many countries due to high availability and relatively low costs [4]. As for China, more than 80% of sludge was dumped improperly, followed by sanitary landfill (13.4%), building materials (2.4%), incineration (0.36%), and land application (0.24%) [5]. However, these techniques are highly controversial due to serious secondary pollution from leachate and air emissions, which is reflected in the significantly high heavy-metal levels found in the vicinity. Therefore, some authors have stressed that traditional wastewater treatment will only shift the problem from water contamination to soil and air pollution, and the potential risks of heavy metals could be greatly reduced, but cannot be entirely eliminated [6]. It is a fact that sewage treatment plants and relevant enterprises are still struggling with rising costs from sludge transport and disposal, and the seeking of the integration planning between environmental protection and sustainable development has become increasingly pronounced.

More recently, attempts have been made to transform environmental waste products into advanced energy conversion and storage materials through direct thermal disposal of sewage sludge [7]. This is particularly crucial for economic and social development since it offers both environmental and energy benefits. Supercapacitors, as new energy storage devices, have garnered great attention in recent years [8]. Carbon-based materials are one of the most widely used electrodes for supercapacitors with desired physical and chemical characterizations [9]. However, they constantly demonstrate lower capacitance and inferior energy density. The use of electrode materials possessing pseudo-capacitance is vital to ensure high capacity as well as high energy and power density [10]. Hence, one of the most significant approaches in the fabrication of high-capacitance materials is the incorporation of transition metals into carbon frameworks. Previous research has demonstrated an obvious enhancement in the electrochemical performance of carbon materials through metal-doping (Mn [11], Ni [12], Co [13], Fe [14]), which can store and release energy reversibly through surface redox reactions and display considerably higher specific capacitance through an additional pseudocapacitive storage contribution between the electro-active materials and the electrolyte [15]. However, expensive metal-doped precursors were used, which makes the synthetic processes costly and difficult to scale-up [16]. Thus, the generation of a low-cost and abundant metal-containing precursor to fabricate high-quality electrode materials is extremely desirable.

Most sewage sludge are naturally rich in carbon (derived from organic polymer flocculants) and heavy metals (e.g., Mn, Cu, Zn, Pb, Cr, and Cd) [17]. Utilizing this heavy metal-contaminated sewage sludge as precursors, metal-doped carbon materials can be prepared on a large scale and at a relatively low cost without any additional metal salts. In addition, some heavy metal oxides can store and release energy reversibly through surface redox reactions between the electro-active materials and the electrolyte, which usually display considerably higher specific capacitance through an additional pseudocapacitive storage contribution [15]. More importantly, the heavy metals would be confined within the carbon frameworks by the complete pyrolysis of sludge, which effectively eliminates the transference of toxic heavy metal. To the best of our knowledge, very few research efforts have been made so far to synthesize metal-doped carbon materials from heavy metal-contaminated sewage sludge. In addition, a number of studies have reported that the preparation and electrochemical performances of electrode materials are sensitive to the composition and structure of the precursor [18]. Generally, the composition of sewage sludge (flocculants and metal) could be easily controlled by the flocculation coefficient during the wastewater treatment process [19]. Therefore, the recycling of waste sludge in an environmentally and economically acceptable way would involve wastewater treatment processes in the early stage.

The main aim of this paper was to study the effects of flocculation conditions on physical properties and pyrolysis products of sludge floc. While dissolved metal pollutants can be removed efficiently from wastewaters by means of the coagulation–flocculation process, they also generate large quantities of toxic sewage sludge. Therefore, we developed a facile, cost-effective, and green approach to synthesize a series of Cu-doped carbon materials through direct carbonization of Cu(II)-containing sludge and their application in energy storage systems (Scheme 1). A starch-based flocculant containing ionizable carboxyl group was synthesized and used as a promising flocculant for heavy metal ion (Cu(II)) flocculation from wastewaters. The effects of flocculant dose and pH were studied for the flocculation of Cu(II) from wastewater using batch studies mode. Sewage sludge with different Cu(II) flocculation capacity were employed as direct precursors for the synthesis of Cu-doped carbon materials. The influence of Cu(II) flocculation capacity on the pyrolysis of sewage sludge and electrochemical performances of electrode material was investigated. The effect of metal content, morphology, and pore structure of the resultant electrode materials on the electrochemical energy storage properties were discussed. Finally, the Cu(II) removal efficiency has also been evaluated relative to possible energy storage benefits, as well as to cost and purification efficiency from wastewater treatment operations.

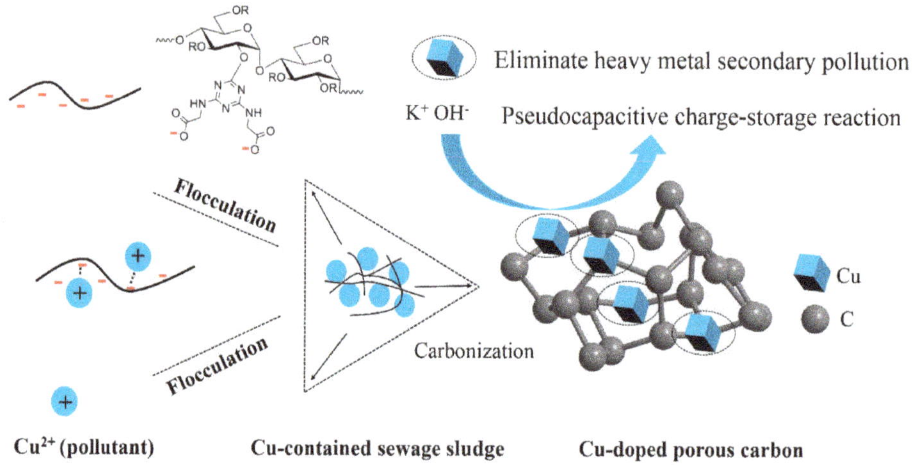

Scheme 1. Fabrication of Cu-doped carbon materials using Cu(II)-containing sewage sludge.

2. Materials and Methods

2.1. Materials

Raw maize starch (food-grade, purity >98%, *w/w*) was purchased from Heng-hui Food Co., Ltd. (Xinjiang, China). Glycine (99% pure), polyvinylidene fluoride (PVDF, 10 wt %), and 1-methyl-2-pyrrolidone were provided by Shanghai Maclean Biochemical Technology Co., Ltd. (Shanghai, China). Ammonium hydroxide (purity >25–28%, *w/w*) and $CuCl_2{\bullet}2H_2O$ (99% pure) were obtained from Fuyu Fine Chemical Co., Ltd. (Tianjin, China). Cyanuric chloride (99% pure) was obtained from the Three Character fine chemical Co., Ltd. (Liaoning, China). Starch-based flocculants (SF) were synthesized using 2-chloro-4,6-diglycino-[1,3,5]-triazine as the etherification agent (Details in Figure S1). Distilled H_2O was used for all experiments.

2.2. Flocculation Experiment

Before each assessment, a flocculant solution of 4 $g{\cdot}L^{-1}$ was obtained by dissolving a certain amount of SF in distilled water. $CuCl_2{\bullet}2H_2O$ was dissolved in distilled water to afford the Cu(II) stock solution of 4 $g{\cdot}L^{-1}$, the pH of which was adjusted to 8.1 using $NH_3{\bullet}H_2O$.

Cu(II) flocculation experiments were performed using batch assessments at room temperature. Specifically, 0.6, 1.4, 1.6, 2, 2.4, and 2.6 mL of flocculant solution were added to a glass beaker (100 mL) containing 1.2 mL Cu(II) stock solution at pH 8.1. After that, water was added to obtain a total volume of 40 mL. Finally, Cu (II) concentration was 120 mg L^{-1}. The suspension was initiated by rapidly stirring at 300 rpm for 5 min, and then slowly mixed for 10 min at 80 rpm. The precipitate was allowed to stand for 30 min and the solution was filtered (2 μm filter paper) to obtain the filtrate for residual concentration (RC) assessments. RC of Cu(II) was estimated from the derived calibration curve (Figure S2). Jar tests were repeated at least three times, the results of which were analyzed with mean values and standard deviation. Furthermore, 2.4 mL flocculant solution was added to a glass beaker (100 mL) containing 1.2 mL Cu(II) stock solution at pH 3, 3.5, 4, 4.5, 5, 5.5, 7, and 8.5, respectively. Water was added to obtain a total volume of 40 mL. After that, the same steps above was used to conduct the remaining flocculation experiments and detection of Cu (II).

Cu(II) removal (R%; Equation (1)) and flocculation capacity (Q; Equation (2)) could be calculated as follows:

$$R = \frac{C_0 - C_f}{C_0} \times 100\% \tag{1}$$

$$Q = V\frac{C_0 - C_f}{m} \tag{2}$$

in which C_0 and C_f ($mg \cdot L^{-1}$) are the initial and final concentrations of Cu(II) in the filtrate, respectively. V (mL) represents the volume of the solution, and m (mg) represents the dried mass of SF. Flocs with different Cu(II) flocculation capacities (Q = 0.21–0.90 $mg \cdot mg^{-1}$, R = 45.1–99.5%) were collected and labeled as SF-x (x = Q = 0.21–0.90 $mg \cdot mg^{-1}$).

2.3. Fabrication of Cu-Doped Carbon Materials Using Cu(II)-Containing Sludge

Cu-doped carbon materials (SFC) were prepared from Cu(II)-containing sludge with various flocculation capacities, i.e., 0.9 (R = 45.0%), 0.6 (R = 99.5%), and 0.25 (R = 45.7%) $mg \cdot mg^{-1}$, respectively. Typically, Cu(II)-containing sludge as a precursor was pyrolyzed (1 h at 300 °C and 2 h at 800 °C) under an Ar atmosphere at a rate of 5 °C min^{-1}. After prolonged cooling down to room temperature, the dark powder produced was dispersed in H_2O while stirring to eliminate ash and other inorganic compounds, and then denoted as SFC-x (x = 0.9, 0.6, and 0.25).

2.4. Characterization Methods

The Cu(II) concentration was determined using Inductively Coupled Plasma-Atomic Emission Spectrometry (Cu: 324.754 nm, ICP-AES, Varian 710E S). The zeta potential (ZP) was assessed using a nanoplus zeta/nano particle analyzer (Otsuka Electronics) and the optical transmittance was determined by means of a UV-visible spectrophotometer (UV-6100S, METASH, Shanghai, China). X-ray diffraction (XRD, Rigaku TTR III) was used for analyzing the sample structure using Cu Kα radiation. Sample morphology was imaged using a scanning electron microscope (SEM, JEOL JSM-6490LV). Transmission electron microscopy (TEM) and high-resolution transmission electron microscopy (HRTEM) were performed using a FEI Tecnai G2 (FEI Company, Hillsboro, OR, USA). X-ray photoelectron spectroscopic (XPS) measurements were conducted on a PHI-5000C (Physical Electronics, Inc. (PHI), Chanhassen, MN, USA). A quantachrome Autosorb surface analyzer (Quantachrome Instruments, Boynton Beach, FL, USA) was used to perform BET surface area measurements at 77.3 K.

2.5. Electrochemical Measurements

Working electrodes were prepared via mixing active materials (80 wt %), carbon (10 wt %), and polyvinylidene fluoride (PVDF, 10 wt %) in 1-methyl-2-pyrrolidone to form a slurry, which was then brushed onto nickel foam (active area: 1 × 1 cm^2) and dried at 60 °C for 24 h. The active material (weight: 5 mg), a Pt foil (1 × 1 cm^2), and saturated calomel electrodes (SCE) were used as working, counter, and reference electrodes, and the electrolyte was 6 M potassium hydroxide (KOH). The electrochemical performance was estimated by cyclic voltammetry (CV), galvanostatic charge/discharge (GCD) measurement, and electrochemical impedance spectroscopy (EIS) on a CHI 760E analyzer (CH Instruments Inc., Shanghai, China). Gravimetric specific capacitance (Csp, $F \cdot g^{-1}$) for single electrodes was calculated from each galvanostatic charge/discharge curve as follows (Equation (3)):

$$Cs = \frac{\mathrm{I} \times \Delta \mathrm{t}}{m \times \Delta V} \tag{3}$$

where m: the mass of the active material, I: discharge current, ΔV: potential change after complete discharge, and Δt: time for complete discharge.

3. Results

3.1. *Removal of Cu(II) from Aqueous Solution*

Visible transmittance change of the flocculant dispersions was recorded for pH values from 3 to 9 at 25 °C (Figure 1a). Clearly, such a starch-based flocculant shows a unique pH-responsive phase transition behavior. As a carboxyl group-functionalized starch derivative, the carboxyl groups get progressively deprotonated at high pH, which enhances the overall hydrophilic property of the flocculant, resulting in a significant increase in solution transmittance [20]. As can be seen from Figure 1a, the flocculant could easily dissolve in water to yield a clear solution with pH ≥8, indicating that a reversible equilibrium was reached with respect to the deprotonation/protonation of the carboxyl group. Zeta potential results also confirmed that the flocculant exhibited a typical anionic property in the investigated pH range of 2.5–9.0 (Figure 1b). The surface charge of the colloidal particles changed from a point of zero charge around pH 2.4 to a highly negative value of −38 mV at pH 8.1. In order to determine the optimum conditions for heavy metal ion removal, the flocculation of Cu(II) was investigated by varying the parameters of flocculation pH and flocculant dosage in simulated wastewater.

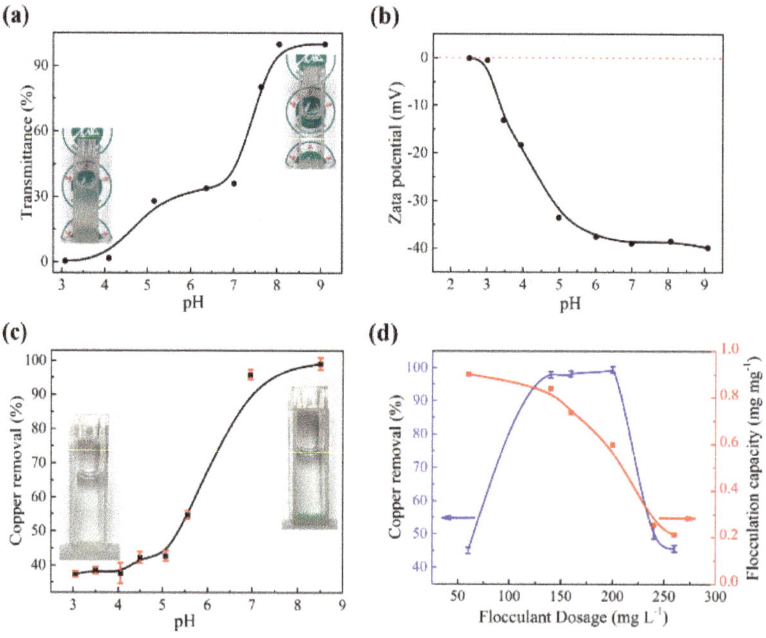

Figure 1. (**a**) pH vs. transmittance of 200 $mg \cdot L^{-1}$ flocculant solution. (**b**) Zeta potential as a function of pH for 200 $mg \cdot L^{-1}$ flocculant solution. (**c**) Influence of pH on % copper removal. (**d**) Effect of flocculant dosage on % copper removal and flocculation capacity.

The influence of pH on the copper removal is depicted in Figure 1c. With an increase in the pH of the flocculation solution, flocculation performance was considerably improved and the Cu(II) removal efficiency reached 99.1% at pH 8.5. As mentioned above, the solubility and zeta potential of the flocculant is greatly affected by the deprotonation of the carboxyl groups and the degree of deprotonation of the carboxyl groups increases with increasing pH [21]. Therefore, heavy metal cations can be eliminated from aqueous media via a combination of charge neutralization and polymer bridging [22].

To investigate the effect of the flocculant dosage on Cu(II) removal, various dosages (60–260 mg· L^{-1}) were exposed to a fixed Cu(II) concentration (120 mg·L^{-1}) at pH 8.1. In general, a low flocculant dosage with simultaneously high heavy metal removal efficiency is greatly desirable for industrial wastewater treatment. An optimal flocculant dosage not only reduces flocculation cost but also decreases the total quantity of sewage sludge, since the flocculant utilization can be reduced to a minimum during wastewater treatment. The effects of flocculant dose on the Cu(II) removal and the corresponding flocculation capacities are shown in Figure 1d. The results signified that Cu(II) removal increased with the increasing dose of flocculant, reached a maximum at the optimal concentration of about 200 mg·L^{-1}, and then decreased with a further increase in dose. The curves were representative of a typical flocculation system that is controlled by charge neutralization mechanisms [21]. Specifically, when the surface charges of the metal cations were completely neutralized by the addition of an anionic flocculant, the maximum Cu(II) removal efficiency was achieved. However, an increase in the Cu(II) residual content at flocculant overdosage could be observed, indicating that the colloid exhibits a re-stabilization phenomenon at higher flocculant doses in the presence of excess anionic charges. Three different Cu-doped carbon materials (SFC-x) were synthesized using flocculation sludge with different Cu(II) flocculation capacity (x = 0.9, 0.6, and 0.25 mg·mg^{-1}).

3.2. Characterization of SFC-x

The structure of SFC-x was assessed by XRD (Figure 2a). The diffraction peaks of SFC-x emerged at 43.5°, 50.4°, and 74.2°, respectively, corresponding to the (111), (200), and (220) planes of Cu (JCPDS file No. 03-1015). Moreover, other characteristic peaks of SFC-0.9 located at 36.4° assigned to the (111) planes of Cu_2O and more peaks of SFC-0.6 were evident, including peaks at 29.5°, 36.4°, 42.3°, and 61.3° assigned to the (110), (111), (200), and (220) planes of Cu_2O, respectively (JCPDS No. 78-2076). These results suggested that Cu-doped carbon materials have been prepared successfully. XPS was employed to investigate the elemental states of SFC-x. Figure 2b shows the XPS survey spectrum of SFC-x; the photoelectron peaks for C 1s (284 eV), N 1s (400 eV), O 1s (531 eV), and Cu 2p (932.6 eV) were observed, respectively. From the XPS data, the Cu content in SFC-0.9 accounted for 21.61 at % of the specimen, which was much higher than that of SFC-0.6 (12.3 at %) and SFC-0.25 (8.5 at %) (Figure S3). These results demonstrated that the Cu content in SFC-x displayed a linear positive correlation with the flocculation capacity of Cu(II). According to a previous study, the electrochemical characteristics of metal-doped electrode materials are closely related to the metal content and metal species [23]. Relatively high metal content means more available active sites were exposed, leading to more redox reactions between the electro-active materials and the electrolyte, and high performance as an anode material for supercapacitors [24]. The deconvolution of the Cu(2p) peaks is shown in Figure 2c. SFC-0.9 shows major peaks of Cu 2p1/2 (952.5 eV) and Cu 2p3/2 (932.7 eV) besides the small peaks (944.20 eV), characteristic of Cu, suggesting that Cu-doped carbon materials have been prepared successfully. The C 1s XPS spectra of SFC-0.9 (Figure 2d) has three peaks of C=C–C (284.7 eV), C–N/C–O (285.6 eV), and C=O (288.1 eV) [25]. The C=C bonds of SFC-x accounted for 35.15 at % (x = 0.9), 36.69 at % (x = 0.6), and 37.07 at % (x = 0.25) of the content (Table S1), respectively, which plays a key role in enhancing the electrochemical performance by improving electron mobility and lowering electrode resistance [26]. It is accepted that the XPS results were in agreement with those of XRD.

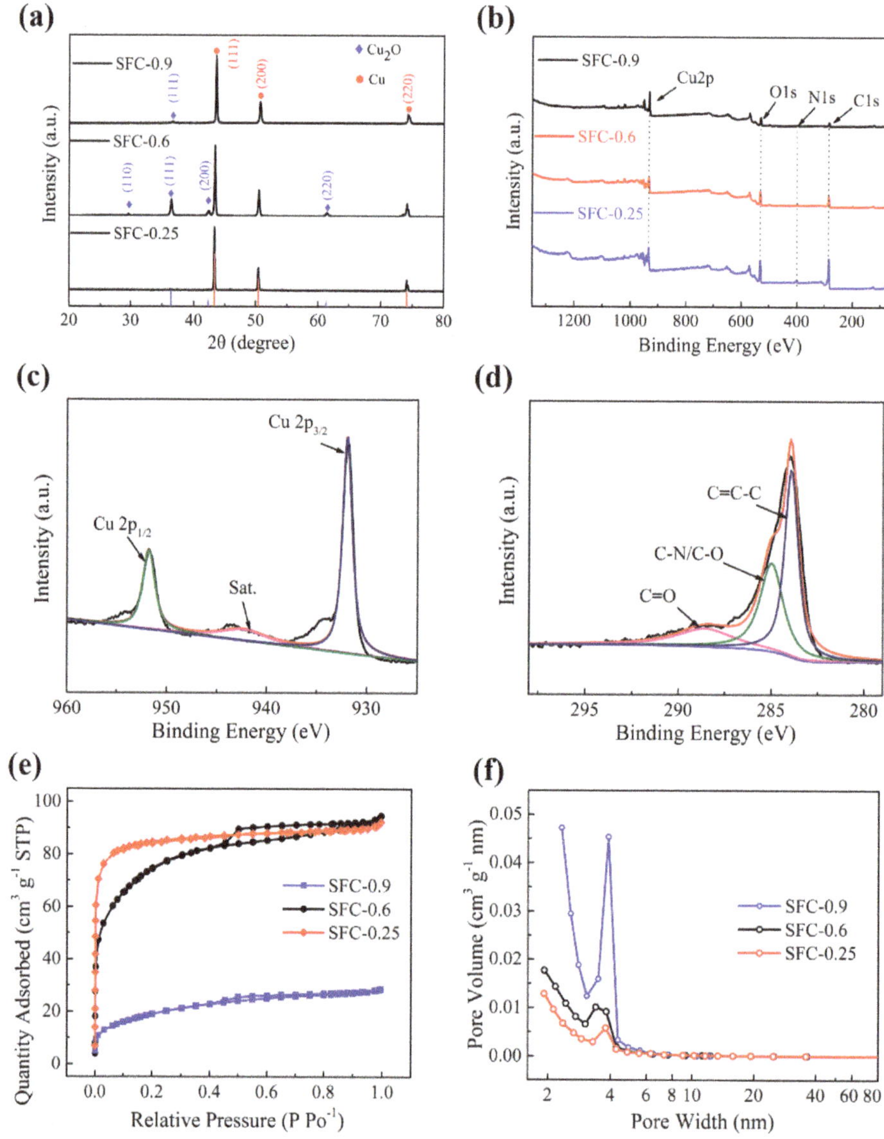

Figure 2. (**a**) XRD patterns of the SFC-x. (**b**) XPS survey spectra of SFC-x. High-resolution XPS spectra of (**c**) C 1s and (**d**) Cu 2p peak of SFC-0.9. (**f**) N2 adsorption-desorption isotherms and (**e**) Barrett-Joyner-Halenda (BJH) pore size distributions of SFC-x.

The BET surface area and pore size of SFC-x were estimated by N_2 adsorption-desorption isotherms and Barrett-Joyner-Halenda (BJH) pore-size distribution analysis. The N_2 adsorption isotherms of SFC-0.9 and SFC-0.6 displayed typical features of type IV isotherms with well-defined plateaus between P/P_0 of 0.1~0.9 and an obvious hysteresis loop at the P/P_0 >0.4 (IUPAC classification), which corresponds to the presence of mesopores (Figure 2e) [27]. Besides, the N_2 adsorption isotherms of SFC-0.25 were type IV isotherms with an unapparent H1-type hysteresis loop [28]. Moreover, the specific surface area (S_{BET}) and pore structure of SFC-x were surprisingly regulated by the flocculation capacity of

Cu(II) in flocs (S_{BET} and pore structure parameters of SFC-x are presented in Table S2). The S_{BET} of SFC-x was enlarged with the decreased copper flocculation capacity. The S_{BET} were 68.54, 258.48, and 285.24 $m^2 \cdot g^{-1}$ for SFC-0.9, SFC-0.6, and SFC-0.25, respectively. Notably, the fact that the SBET of SFC-0.9 was the smallest may be ascribed to the carbon content of SFC-0.9, which was the lowest (Figure 2b). The pore size distribution of SFC-x was assessed by means of the BJH model on the adsorption isotherm branches (Figure 2f). The pore size distributions of all samples were mainly in the range of 2 to 6 nm, which corresponded to the mesopore and was beneficial to the specific capacity via decreasing the ion transfer impedance from the electrolyte to micropores and inducing the electrical double layer formation [29]. Based on the suitable pore size distribution, SFC-x is expected to have superior capacitive performance.

Morphological features of SFC-x were analyzed through SEM and HRTEM. It was clear that the sizes and morphologies of SFC-x were dependent upon the Cu(II) flocculation capacity (heavy metal content) in flocs. As shown in Figure 3a, the SFC-0.6 has a cubic-like structure synthesized with Cu(II) flocculation capacity of 0.6 $mg \cdot mg^{-1}$ in flocs. A previous study has shown that porous Cu_2O–Cu cubes can be prepared by reducing Cu(II) chelate at higher temperatures [30]. We also found that, as the Cu(II) flocculation capacity increased from 0.6 to 0.9 $mg \cdot mg^{-1}$ in sludge, the morphologies of SFC-0.9 changed from cubic to accumulated nano-cubes particles (Figure S4a). In contrast, for SFC-0.25, we observed a change in morphology from cubic to a hierarchical flower-like structure with increasing Cu(II) flocculation capacity from 0.25 to 0.6 $mg \cdot mg^{-1}$ (Figure S4b). The results indicated that the morphology of Cu-doped carbon materials can be easily tuned by the Cu(II) flocculation capacity in sludge.

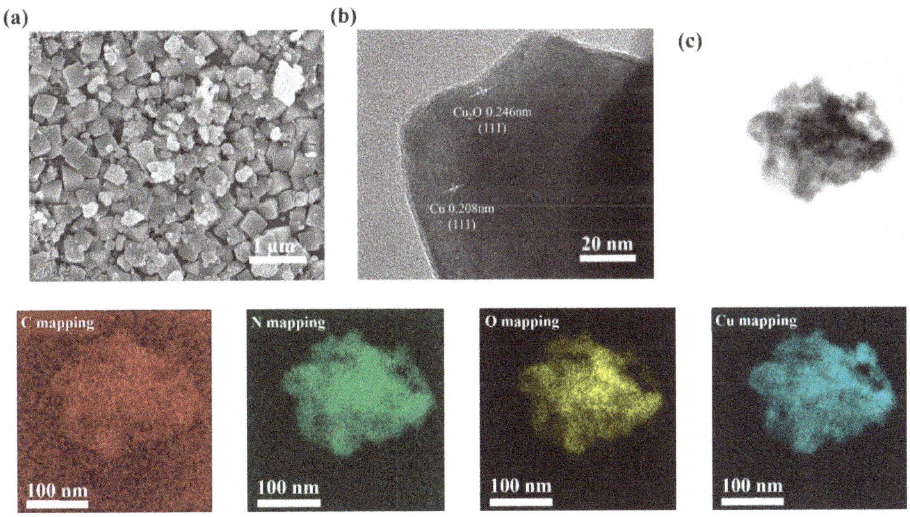

Figure 3. (**a**) SEM images, (**b**) HRTEM, and (**c**) TEM-mapping images of SFC-0.6.

The microstructure of the Cu-doped carbon materials was further investigated by HRTEM technique. From the insert in Figure 3b, lattice spacings of SFC-0.6 emerged at 0.208 nm and 0.246 nm, corresponding to the (111) crystal plane of Cu and the (111) plane of Cu_2O. In addition, similar lattice fringes have also been observed from HRTEM image of SFC-0.9 and SFC-0.25 (Figure S4b,d), with average size of lattice spacings of about 0.208 nm, which corresponds to the (111) lattice planes of the Cu structure. These results were consistent with the peaks of XRD, further indicating that Cu-doped carbon materials have been prepared successfully. Additionally, the elemental mapping images of SFC-0.6 revealed that the C, N, O, and Cu elements were uniformly dispersed (Figure 3c). As mentioned above, the main flocculation mechanism of Cu(II) was proposed on the basis of charge neutralization mechanisms, thus, highly dispersed Cu nanoparticles on the carbon framework were attained by combining the coagulation of Cu(II) and thermal treatment of the flocs. Uniform distribution of the Cu nanoparticles were easily accessed by the electrolyte ions as an electrode material, resulting in more activity towards redox reactions for supercapacitors [31].

3.3. Electrochemical Performance of SFC-x

The electrochemical performance of SFC-x was assessed via the three-electrode configuration in 6 M KOH. Figure 4a displays the CV curves of SFC-x at a potential scan rate of 5 mV·s^{-1} (potential window: −1 to 0 V). It was apparent that the area surrounded by the CV curves of the SFC-0.9 was larger than those of the SFC-0.6 and SFC-0.25 at the same scan rate, signifying that the SFC-0.9 possessed the highest specific capacitance. Besides, the curve shape had similarities and discrepancies from the rectangular shape controlled by the electrical double layer capacitance. The clear current peaks at −0.39 V and −0.12 V were ascribed to the oxidation of Cu^0 to Cu^{2+} in the electrode [32]. In addition, two peak currents were evident when scanning from 0 V to −1 V (at −0.38 V and −0.82 V), ascribed to the reduction of Cu^{2+} to Cu^0 in the electrode [33]. These results demonstrated that the pseudo-capacitance behavior of SFC-x was due to the transformations between Cu^0 and Cu^{2+}. From the redox peaks of copper in aqueous KOH electrolyte [34,35], the pseudo-capacitance behavior in the CV curves was associated with the following reactions (Equations (4) and (5)):

$$Cu_2O + 2OH^- \leftrightarrow 2\text{CuO} + H_2O + 2e^- \tag{4}$$

$$2Cu + 2OH^- \leftrightarrow Cu_2O + H_2O + 2e^- \tag{5}$$

The CV measurements of SFC-0.6 under different scan rates are presented in Figure 4b. It was observed that the shape of the CV curves of SFC-0.6 was not obviously distorted with an increase in scan rates, indicating an ideal capacitive behavior. Obviously, the intensity of current peaks decreased upon increasing the scan rate from 5 to 50 mV s^{-1}. Since the ions do not have sufficient time to diffuse into the minute pores and reach the active sites at higher scanning rates, which makes a very weak redox cycling in the electrode surface, thus leading to a decline in the current peaks. Furthermore, the GCD curves are a powerful measure for evaluation of supercapacitive performance. From GCD measurements of SFC-x at a current density of 1 A g^{-1} (Figure 4c), the specific capacitances of SFC-0.9, SFC-0.6, and SFC-0.25 were 637.9, 389.9, and 308.6 F·g^{-1}, respectively. These results indicated that the flocculation capacity of Cu(II) in flocs was linearly positively related to the specific capacitance values of SFC-x. Moreover, it was evident that changes occurred twice in the slope of the potential in the discharge and charge curves, e.g., the slope changes occurred during S1, S2, S3, and S4 for SFC-0.9, which correlated with the current peaks observed in the CV curves. Compared with the ideal smooth charge/discharge curve, the charge/discharge curves of SFC-x display deviation further indicated that the SFC-x exhibits pseudo-capacitive behaviors. The GCD measurements of SFC-0.6 at various current densities are shown in Figure 4d. The specific capacitance values of SFC-0.6 were 427.1, 389.9, 346.9, and 187.4 F·g^{-1} that corresponded to the current densities of 0.5, 1, 2, and 5 A·g^{-1}, respectively. Notably, the shape of the GCD curve was maintained up to 5 A·g^{-1}, indicating a low equivalent series resistance [36]. A high capacitance of 187.4 F·g^{-1} could still be retained at a high current density of 5 A· g^{-1}, which was greater than that of other Cu-based nanocomposites [37,38].

As shown in the Nyquist plot in Figure 5a, the spectra of the prepared samples were composed of a high-middle frequency region (semi-circle) associated with the charge-transfer at the electrode/electrolyte interface, and a low-frequency region (straight line) due to Warburg impedance [39]. The equivalent series resistance (ESR) of the SFC-x was acquired from the X-intercept of the sloping line at the low frequency region. Compared to SFC-0.6 and SFC-0.25, there was a maximum tilt angle line for SFC-0.9, indicating the lowest ESR for the latter. These results revealed that the conductivity and charge transfer efficiency of SFC–x could be significantly enhanced via augmenting the flocculation capacity of Cu(II). The charge transfer resistance (R_{CT}) values of the electrodes can be represented by the radius of semicircle arcs on the x-axis, and a smaller semicircle represents a smaller charge transfer resistance [40]. It was also found that the R of SFC-0.9 was clearly lower than that of the others, indicating a fast charge transfer. The low resistance was chiefly attributed to the high conductivity and low resistance metal–metal contacts.

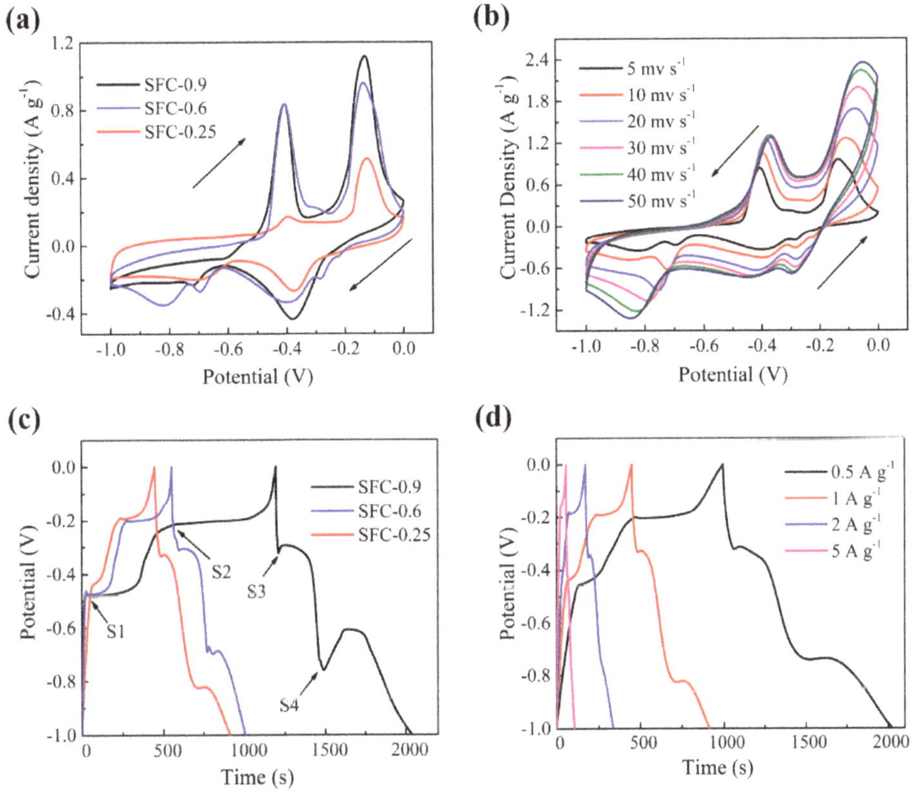

Figure 4. (**a**) Cyclic voltammograms (CV) of SFC-x in a three-electrode configuration (scan rate = 5 mV· s^{-1}). (**b**) Cyclic voltammograms of SFC-0.6 at various scan rates (5–50 mV·s^{-1}). (**c**) Galvanostatic charge/discharge (GCD) curves of SFC-x in a three-electrode configuration (current densities = 1 A·g^{-1}). (**d**) Galvanostatic charge/discharge curves of SFC-0.6 at various current densities (0.5–5 A·g^{-1}).

The specific capacitance of SFC-x as a function of current density is shown in Figure 5b (Calculated from GCD of SFC-x in curves of Figure 4d and Figure S5). The specific capacitance values of SFC-x decreased rapidly with an increase in current density, which could be attributed to the decrease in IR and the difficulty of the electrolyte ions in accessing the reactive sites of SFC-x for redox reactions at high current densities. The capacitance values of SFC-x were similar to, or greater than that of previously published reports on Cu nanoparticles [32], $Cu_2O/CuO/RGO$ [41], 3D porous CuO [42], and Cu_2O/RGO [43]. Since the electrode materials were prepared from toxic sewage sludge, a byproduct of wastewater treatments, this process is crucial for economic and social development because it is beneficial with respect to both environmental and financial aspects.

Although the specific capacitance of SFC-0.9 was the highest among the samples, it has the lowest Cu(II) removal rate (R = 45.02%) which might have a detrimental impact on the environment. Figure 5c presents the Cu(II) removal efficiency and the possible energy storage benefits. With an increasing flocculant dosage, Cu(II) removal increased and the environmental impact was decreased, but there was a rapid decrease in Cu(II) removal with a further increase in flocculant dosage due to the re-stabilization of flocs. However, as a result of the low content of Cu supported on carbon materials, electrochemical performance of SFC-x gradually decreased with the increasing flocculant dosage. From a practical point of view, equilibrium should exist between environmental impact (heavy metals removal efficiency from wastewater) and energy (specific capacitance as a supercapacitor electrode).

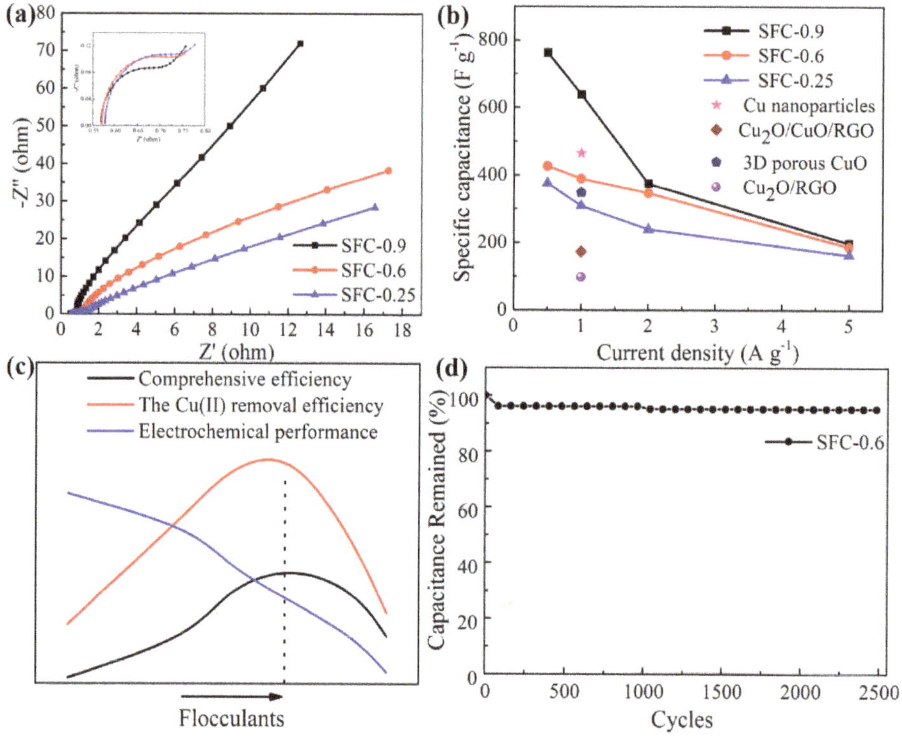

Figure 5. (**a**) Nyquist plots of SFC-x in 6 M KOH electrolyte. (**b**) The specific capacitance of various electrodes as a function of current density based on the GCD curves. (**c**) The Cu(II) removal efficiency and the possible energy storage benefits. (**d**) Cycling performance of SFC-0.6 at a current density of 5 A·g^{-1}.

These results suggested that the flocculant dosage up to 200 mg·L^{-1} is very promising in terms of comprehensive efficiency, at which more than 99% Cu(II) removal efficiency could be achieved, while the resulting annealed products (SFC-0.6) exhibited a high specific capacity (389.9 F·g^{-1} at 1 A·g^{-1}). Also, long cycle life is an essential requirement of an electrode material for practical application in supercapacitors. The cycling performance of SFC-0.6 was assessed at high current densities of 5 A·g^{-1} for a 2500 charge-discharge cycle (Figure 5d). As expected, SFC-0.6 displayed high stability with only a 4% decline in specific capacitance after 2500 cycles. The high cycling stability could be ascribed to the cube-like structure and the large surface area of the porous structure, which mitigated the volume expansion during repeated charge/discharge phases. Two-electrode test cells are more closely related to the physical configuration and charge transfer for practical application of a supercapacitor, though lower values of capacitance are typically observed [44]. The electrochemical performance of SFC-0.6 was estimated from the GCD curves in a two-electrode system (Figure S6). An energy density as high as 7.6 Wh·kg^{-1} (at 0.5 A·g^{-1}) was obtained, with a power density greater than 3400 W·kg^{-1} at a current load of 10 A·g^{-1}. The superior supercapacitor performances of SFC-x were attributed to high Cu content supported on carbon, as well as the fully developed pore structures. Furthermore, the formation of Cu-doped carbon materials contributed to redox pseudo-capacitance during rapid charge and discharge cycles, which further enhanced the electrochemical performance of the electrode for supercapacitors. More importantly, the heavy metal would be confined within the carbon matrix by the complete pyrolysis of sludge, which effectively prevents the secondary heavy metal pollution and emissions from the leachate (Scheme 1).

4. Conclusions

In this work, a carboxyl group-functionalized starch derivative was synthesized and used as an effective flocculant for Cu(II) removal from wastewaters. Employing these Cu-contaminated sewage sludge as precursors, Cu-doped carbon materials were prepared as efficient electrode materials for supercapacitors through one-step carbonization without any additional metal salt. The results revealed that the specific capacitance of the resulting annealed products was linearly positively correlated to the Cu(II) flocculation capacity. With respect to the environmental capacity and energy capacity, the Cu(II) removal efficiency has been analyzed and compared to the possible energy benefits. The flocculant dosage up to 200 mg·L^{-1} was an equilibrium point existing between environmental impact and energy, as high as 99.50% Cu(II) removal efficiency could be achieved. Moreover, the resulting annealed product (SFC-0.6) exhibited a high specific capacity (389.9 F·g^{-1} at 1 A·g^{-1}) and long cycling stability, with only 4% loss after 2500 cycles. This work presents a new approach to recycling heavy metal-contaminated sewage sludge to synthesize advanced energy storage materials, which is highly promising for commercial applications ranging from the energy to environment fields.

Supplementary Materials: The following are available online at http://www.mdpi.com/2079-4991/9/6/892/s1. Figure S1. Starch-based flocculant containing ionizable carboxyl group was synthesized using 2-chloro-4,6-diglycino-[1,3,5]-triazine (CDT) as etherification agent, Figure S2. Standard curve of Copper, Figure S3. The elemental analysis of SFC-x, Table S1. Different types of carbon bonds, Table S2. Physical parameters for SFC-x, Figure S4. (a) SEM images and (b) HRTEM of SFC-0.9. (c) SEM images of and (d) HRTEM of SFC-0.25, Figure S5. GCD curves of (a) SFC-0.9 and (b) SFC-0.25 electrode at different current densities in three-electrode, Figure S6. (a) CV of SFC-0.6 at various scan rates (5–50 mV s^{-1}) in two-electrode (b) GCD curves of SFC-0.6 electrode at various current densities (0.5–10 A g^{-1}) in two-electrode. (c) The specific capacitance of various electrodes as a function of current density based on the GCD curves. (d) Energy density with respect to the power density of the SFC-0.6.

Author Contributions: Z.T., F.Y., and Y.S. designed the experiments. Y.S., F.Y., and X.G. administered the experiments. Z.T. and L.L. performed experiments. Z.T. and Y.S. collected data. L.C., Y.L., and Y.X. gave conceptual advice. All authors analyzed, discussed the data, and wrote the manuscript.

Funding: This research was funded by National Natural Science Foundation of China (51764049, 21467024 and 21661027), International Scientific and Technological Cooperation Project of Xinjiang Bingtuan (2018BC001), Transformation of Technological Achievements in Shihezi University (CGZH201715) and the project was funded by Scientific Research Start-up Fund for High-Level Talents, Shihezi University (KX0138).

Conflicts of Interest: The authors declare no conflict of interest.

References

1. Xue, Y.; Wang, C.; Hu, Z.; Zhou, Y.; Xiao, Y.; Wang, T. Pyrolysis of sewage sludge by electromagnetic induction: Biochar properties and application in adsorption removal of Pb(II), Cd(II) from aqueous solution. *Waste Manag.* **2019**, *89*, 48–56. [CrossRef] [PubMed]
2. Lu, K.; Yang, X.; Gielen, G.; Bolan, N.; Ok, Y.S.; Niazi, N.K.; Xu, S.; Yuan, G.; Chen, X.; Zhang, X.; et al. Effect of bamboo and rice straw biochars on the mobility and redistribution of heavy metals (Cd, Cu, Pb and Zn) in contaminated soil. *J. Environ. Manag.* **2017**, *186*, 285–292. [CrossRef] [PubMed]
3. Singh, R.P.; Agrawal, M. Potential benefits and risks of land application of sewage sludge. *Waste Manag.* **2008**, *28*, 347–358. [CrossRef] [PubMed]
4. Kacprzak, M.; Neczaj, E.; Fijałkowski, K.; Grobelak, A.; Grosser, A.; Worwag, M.; Rorat, A.; Brattebo, H.; Almås, Å.; Singh, B.R. Sewage sludge disposal strategies for sustainable development. *Environ. Res.* **2017**, *156*, 39–46. [CrossRef] [PubMed]

5. Yang, G.; Zhang, G.; Wang, H. Current state of sludge production, management, treatment and disposal in China. *Water Res.* **2015**, *78*, 60–73. [CrossRef] [PubMed]
6. Zhu, X.-N.; Nie, C.-C.; Zhang, H.; Lyu, X.-J.; Qiu, J.; Li, L. Recovery of metals in waste printed circuit boards by flotation technology with soap collector prepared by waste oil through saponification. *Waste Manag.* **2019**, *89*, 21–26. [CrossRef] [PubMed]
7. Chen, W.; Wang, X.; Luo, M.; Yang, P.; Zhou, X. Fast one-pot microwave preparation and plasma modification of porous carbon from waste lignin for energy storage application. *Waste Manag.* **2019**, *89*, 129–140. [CrossRef]
8. Tran, C.; Lawrence, D.; Richey, F.W.; Dillard, C.; Elabd, Y.A.; Kalra, V. Binder-free three-dimensional high energy density electrodes for ionic-liquid supercapacitors. *Chem. Commun.* **2015**, *51*, 13760–13763. [CrossRef]
9. Borenstein, A.; Hanna, O.; Attias, R.; Luski, S.; Brousse, T.; Aurbach, D. Carbon-based composite materials for supercapacitor electrodes: A review. *J. Mater. Chem. A* **2017**, *5*, 12653–12672. [CrossRef]
10. Wang, X.; Shen, G. Intercalation pseudo-capacitive $TiNb_2O_7$@carbon electrode for high-performance lithium ion hybrid electrochemical supercapacitors with ultrahigh energy density. *Nano Energy* **2015**, *15*, 104–115. [CrossRef]
11. Li, J.; Ren, Y.; Wang, S.; Ren, Z.; Yu, J. Transition metal doped MnO_2 nanosheets grown on internal surface of macroporous carbon for supercapacitors and oxygen reduction reaction electrocatalysts. *Appl. Mater. Today* **2016**, *3*, 63–72. [CrossRef]
12. Tian, D.; Lu, X.; Zhu, Y.; Li, M.; Wang, C. Fabrication of two-dimensional metal-organic frameworks on electrospun nanofibers and their derived metal doped carbon nanofibers for an advanced asymmetric supercapacitor with a high energy density. *J. Power Sources* **2019**, *413*, 50–58. [CrossRef]
13. Wu, Z.; Winter, A.; Chen, L.; Sun, Y.; Turchanin, A.; Feng, X.; Müllen, K. Three-dimensional nitrogen and boron co-doped graphene for high-performance all-solid-state supercapacitors. *Adv. Mater.* **2012**, *24*, 5130–5135. [CrossRef] [PubMed]
14. Liu, L.; Zhang, H.; Fang, L.; Mu, Y.; Wang, Y. Facile preparation of novel dandelion-like Fe-doped $NiCo_2O_4$ microspheres@nanomeshes for excellent capacitive property in asymmetric supercapacitors. *J. Power Sources* **2016**, *327*, 135–144. [CrossRef]
15. Shen, L.; Yu, L.; Wu, H.B.; Yu, X.-Y.; Zhang, X.; Lou, X.W. Formation of nickel cobalt sulfide ball-in-ball hollow spheres with enhanced electrochemical pseudocapacitive properties. *Nat. Commun.* **2015**, *6*, 6694. [CrossRef] [PubMed]
16. Shang, L.; Yu, H.; Huang, X.; Bian, T.; Shi, R.; Zhao, Y.; Waterhouse, G.I.N.; Wu, L.; Tung, C.; Zhang, T. Well-Dispersed ZIF-Derived Co,N-Co-doped carbon nanoframes through mesoporous-silica-protected calcination as efficient oxygen reduction electrocatalysts. *Adv. Mater.* **2016**, *28*, 1668–1674. [CrossRef]
17. Uddin, M.K. A review on the adsorption of heavy metals by clay minerals, with special focus on the past decade. *Chem. Eng. J.* **2017**, *308*, 438–462. [CrossRef]
18. Behm, M.; Irvine, J.T.S. Influence of structure and composition upon performance of tin phosphate based negative electrodes for lithium batteries. *Electrochim. Acta* **2002**, *47*, 1727–1738. [CrossRef]
19. Guibal, E.; Roussy, J. Coagulation and flocculation of dye-containing solutions using a biopolymer (Chitosan). *React. Funct. Polym.* **2007**, *67*, 33–42. [CrossRef]
20. Jia, S.; Yang, Z.; Yang, W.; Zhang, T.; Zhang, S.; Yang, X.; Dong, Y.; Wu, J.; Wang, Y. Removal of Cu(II) and tetracycline using an aromatic rings-functionalized chitosan-based flocculant: Enhanced interaction between the flocculant and the antibiotic. *Chem. Eng. J.* **2016**, *283*, 495–503. [CrossRef]
21. Shi, Y.; Ju, B.; Zhang, S. Flocculation behavior of a new recyclable flocculant based on pH responsive tertiary amine starch ether. *Carbohydr. Polym.* **2012**, *88*, 132–138. [CrossRef]
22. Fu, F.; Wang, Q. Removal of heavy metal ions from wastewaters: A review. *J. Environ. Manag.* **2011**, *92*, 407–418. [CrossRef]
23. Giovanni, M.; Poh, H.L.; Ambrosi, A.; Zhao, G.; Sofer, Z.; Šaněk, F.; Khezri, B.; Webster, R.D.; Pumera, M. Noble metal (Pd, Ru, Rh, Pt, Au, Ag) doped graphene hybrids for electrocatalysis. *Nanoscale* **2012**, *4*, 5002–5008. [CrossRef] [PubMed]
24. Vangari, M.; Pryor, T.; Jiang, L. Supercapacitors: Review of materials and fabrication methods. *J. Energy Eng.-ASCE* **2013**, *139*, 72–79. [CrossRef]
25. Yu, M.; Zhang, J.; Li, S.; Meng, Y.; Liu, J. Three-dimensional nitrogen doped holey reduced graphene oxide framework as metal-free counter electrodes for high performance dye-sensitized solar cells. *J. Power Sources* **2016**, *308*, 44–51. [CrossRef]

26. Teo, E.Y.L.; Muniandy, L.; Ng, E.-P.; Adam, F.; Mohamed, A.R.; Jose, R.; Chong, K.F. High surface area activated carbon from rice husk as a high performance supercapacitor electrode. *Electrochim. Acta* **2016**, *192*, 110–119. [CrossRef]
27. Chen, D.; Li, L.; Xi, Y.; Li, J.; Lu, M.; Cao, J.; Han, W. Self-assembly of biomass microfibers into 3D layer-stacking hierarchical porous carbon for high performance supercapacitors. *Electrochim. Acta* **2018**, *286*, 264–270. [CrossRef]
28. Shi, R.; Han, C.; Li, H.; Xu, L.; Zhang, T.; Li, J.; Lin, Z.; Wong, C.; Kang, F.; Li, B. NaCl-templated synthesis of hierarchical porous carbon with extremely large specific surface area and improved graphitization degree for high energy density lithium ion capacitors. *J. Mater. Chem. A* **2018**, *6*, 17057–17066. [CrossRef]
29. Dong, C.; Wang, Y.; Xu, J.; Cheng, G.; Yang, W.; Kou, T.; Zhang, Z.; Ding, Y. 3D binder-free Cu_2O@Cu nanoneedle arrays for high-performance asymmetric supercapacitors. *J. Mater. Chem. A* **2014**, *2*, 18229–18235. [CrossRef]
30. Sun, S. Recent advances in hybrid Cu_2O-based heterogeneous nanostructures. *Nanoscale* **2015**, *7*, 10850–10882. [CrossRef]
31. Liu, J.; Jiang, J.; Cheng, C.; Li, H.; Zhang, J.; Gong, H.; Fan, H.J. Co_3O_4 nanowire@mnO2 ultrathin nanosheet core/shell arrays: A new class of high-performance pseudocapacitive materials. *Adv. Mater.* **2011**, *23*, 2076–2081. [CrossRef]
32. Teng, S.; Siegel, G.; Prestgard, M.C.; Wang, W.; Tiwari, A. Synthesis and characterization of copper-infiltrated carbonized wood monoliths for supercapacitor electrodes. *Electrochim. Acta* **2015**, *161*, 343–350. [CrossRef]
33. Li, H.; Chen, Z.; Wang, Y.; Zhang, J.; Yan, X.; Lei, T.; Peng, Z.; Zhou, Z.; Lv, H. Facile preparation of S-doped Cu/C core-shell composite for high-performance anode of pseudocapacitors. *Electrochim. Acta* **2018**, *259*, 1030–1036. [CrossRef]
34. Vidhyadharan, B.; Misnon, I.I.; Aziz, R.A.; Padmasree, K.P.; Yusoff, M.M.; Jose, R. Superior supercapacitive performance in electrospun copper oxide nanowire electrodes. *J. Mater. Chem. A* **2014**, *2*, 6578–6588. [CrossRef]
35. Wang, G.; Huang, J.; Chen, S.; Gao, Y.; Cao, D. Preparation and supercapacitance of CuO nanosheet arrays grown on nickel foam. *J. Power Sources* **2011**, *196*, 5756–5760. [CrossRef]
36. Xiao, X.; Liu, X.; Zhao, H.; Chen, D.; Liu, F.; Xiang, J.; Hu, Z.; Li, Y. Facile shape control of Co_3O_4 and the effect of the crystal plane on electrochemical performance. *Adv. Mater.* **2012**, *24*, 5762–5766. [CrossRef] [PubMed]
37. Pendashteh, A.; Mousavi, M.F.; Rahmanifar, M.S. Fabrication of anchored copper oxide nanoparticles on graphene oxide nanosheets via an electrostatic coprecipitation and its application as supercapacitor. *Electrochim. Acta* **2013**, *88*, 347–357. [CrossRef]
38. Chen, H.; Zhou, M.; Wang, T.; Li, F.; Zhang, Y.X. Construction of unique cupric oxide–manganese dioxide core–shell arrays on a copper grid for high-performance supercapacitors. *J. Mater. Chem. A* **2016**, *4*, 10786–10793. [CrossRef]
39. Nara, H.; Morita, K.; Mukoyama, D.; Yokoshima, T.; Momma, T.; Osaka, T. Impedance analysis of $LiNi_{1/3}Mn_{1/3}Co_{1/3}O_2$ cathodes with different secondary-particle size distribution in Lithium-ion Battery. *Electrochim. Acta* **2017**, *241*, 323–330. [CrossRef]
40. Xu, B.; Hou, S.; Cao, G.; Wu, F.; Yang, Y. Sustainable nitrogen-doped porous carbon with high surface areas prepared from gelatin for supercapacitors. *J. Mater. Chem. A* **2012**, *22*, 19088–19093. [CrossRef]
41. Wang, K.; Dong, X.; Zhao, C.; Qian, X.; Xu, Y. Facile synthesis of Cu_2O/CuO/RGO nanocomposite and its superior cyclability in supercapacitor. *Electrochim. Acta* **2015**, *152*, 433–442. [CrossRef]
42. Dong, X.; Wang, K.; Zhao, C.; Qian, X.; Chen, S.; Li, Z.; Liu, H.; Dou, S. Direct synthesis of RGO/Cu_2O composite films on Cu foil for supercapacitors. *J. Alloy. Compd.* **2014**, *586*, 745–753. [CrossRef]
43. Yu, L.; Jin, Y.; Li, L.; Ma, J.; Wang, G.; Geng, B.; Zhang, X. 3D porous gear-like copper oxide and their high electrochemical performance as supercapacitors. *CrystEngComm* **2013**, *15*, 7657–7662. [CrossRef]
44. Stoller, M.D.; Ruoff, R.S. Best practice methods for determining an electrode material's performance for ultracapacitors. *Energy Environ. Sci.* **2010**, *3*, 1294–1301. [CrossRef]

© 2019 by the authors. Licensee MDPI, Basel, Switzerland. This article is an open access article distributed under the terms and conditions of the Creative Commons Attribution (CC BY) license (http://creativecommons.org/licenses/by/4.0/).

nanomaterials

MDPI

Article

Mesopore-Rich Activated Carbons for Electrical Double-Layer Capacitors by Optimal Activation Condition

Hye-Min Lee [1,2], Kay-Hyeok An [3,*], Soo-Jin Park [2,*] and Byung-Joo Kim [1,*]

1 Research Center for Carbon Convergence Materials, Korea Institute of Carbon Convergence Technology, Jeonju 54852, Korea; hmlee2014@hanmail.net
2 Department of Chemistry, Inha University, Incheon 22212, Korea
3 Department of Nano & Advanced Materials Engineering, Jeonju University, Jeonju 55069, Korea
* Correspondence: khandragon@jj.ac.kr (K.-H.A.); sjpark@inha.ac.kr (S.-J.P.); kimbj2015@gmail.com (B.-J.K.); Tel.: +82-63-219-3710 (B.-J.K.)

Received: 18 March 2019; Accepted: 10 April 2019; Published: 12 April 2019

Abstract: In this study, activated polymer-based hard carbon using steam activation (APHS) with mesopore-rich pore structures were prepared for application as electrodes in electrical double-layer capacitors (EDLC). The surface morphologies and structural characteristics of APHS were observed using scanning electron microscopy and X-ray diffraction analysis, respectively. The textural properties were described using Brunauer-Emmett-Teller and Barrett-Joyner-Halenda equations with N_2/77 K adsorption isotherms. APHS were prepared under various steam activation conditions to find optimal ones, which were then applied as electrode materials for the EDLC. The observed specific surface areas and total pore volumes of the APHS were in the range 1170–2410 m^2/g and 0.48–1.22 cm^3/g, respectively. It was observed that pore size distribution mainly depended on the activation time and temperature, and that the volume of pores with size of 1.5–2.5 nm was found to be a key factor determining the electrochemical capacity.

Keywords: hard carbon; steam activation; polymeric precursor; electrical double-layer capacitor

1. Introduction

Electrical double-layer capacitors (EDLCs) are very attractive for use in potential energy storage devices because of their high power density, quick charge-discharge rate, and long maintenance-free operation life [1–3]. Energy storage behavior of an EDLC arises mainly from the separation of electronic and ionic charges at the interface between the electrode materials and the electrolyte solution [4]. Therefore, the electrochemical behavior of an EDLC is determined by the textural properties of the active material. Activated carbons (AC) have excellent textural properties and high electrical conductivity, which have made them the most widely used active materials for EDLCs [5–7].

Many studies have focused on the specific surface area among the textural properties of AC [5,6]; however, Baek's research showed that different pore structures are also required, depending on the salt and solvent of the electrolyte [7]. In conclusion, to improve the electrochemical properties of EDLCs, it is necessary to optimize the pore structure of the AC for each electrolyte.

The pore structure of AC is known to be greatly influenced by the precursor [8–10], activation method [11–14], and pyrolysis conditions [15,16]. The final pore size distribution is mainly determined by the activation method. ACs are prepared by physical activation (gasification of a char in oxidizing gases) or by chemical activation (pyrolysis of precursor impregnated with chemical reagents).

Chemical activation is generally done by mixing carbonaceous materials with chemical activating agents (KOH, H_3PO_4, $ZnCl_2$, etc.), followed by pyrolysis at 400–900 °C [17–19]. This process gives rise to

AC with a high specific surface area (>2000 m^2/g) mainly of micropores, with some sub-mesopores [5,17]. Such a high specific surface area is ascribed to partial gasification and expansion of the interlayer spacing between crystallites through simultaneous intercalation and deintercalation [17]. However, this approach has disadvantages that include corrosiveness of the chemical agents and the washing process that is necessary to remove the chemical agents.

Physical activation is done by carbonization of carbon precursors in an inert atmosphere to remove non-carbon elements, followed by activation in the presence of suitable oxidizing gasifying agents (O_2, CO_2, or H_2O) to develop the porosity, usually in the temperature range 600–1200 °C [14–16]. Pores are formed by the oxidation of crystallites by physical activation, during which the size of the crystallite affects pore characteristics as much as the activation method does [16]. Baek et al., reported that hard carbon (HC) of low crystallinity could be used to obtain AC of high specific surface area using physical activation [14,15].

In general, the textural properties (specific surface area, pore volume) of AC resulting from chemical activation are known to be better than those from physical activation [17]. This is why most AC studies for EDLC focus on chemical activation methods. However, the process cost of chemical activation is much higher than that of physical activation. Thus, research is needed to produce AC with excellent pore characteristics through physical activation.

Most commercial grade AC is derived from naturally occurring precursors such as wood [19], coal [20], and coconut shells [21]. However, naturally occurring precursors contain large quantities of ash [22]. For the production of AC, ash causes problems such as capacity reduction, gas creation, and swelling of the EDLC [23,24]. Therefore, AC produced from naturally occurring precursors requires a separate ash removal process.

Polymeric precursors have structural features similar to those in coal, but contain many fewer mineral impurities (from catalysts), which can be controlled to very low levels during their synthesis [25]. However, polymer-based AC with high specific surface area has been reported as a result of chemical activation [17,24]. Generally, a polymer-based precursor with high carbonization yield has high crystallinity; therefore, it is difficult to produce AC having a high specific surface area by physical activation [26].

In this work, activated polymer-based hard carbon using steam activation (APHS) with a high specific surface area and mesopore-rich pore structure was prepared from polymeric precursors with low crystallinity. The pore structure of the APHS obtained was studied using N_2 adsorption. The APHS thus prepared was applied as electrodes for the EDLC, and its specific capacitance was discussed in relation to the pore structure.

2. Experiment Details

2.1. Materials

Polymeric precursor (polyurethane) was obtained from Aekyung Petrochemical Co., Ltd. (Wanju, Korea) [27]. Portions of the polymeric precursor (5 g) were heated to 900 °C at 10 °C/min in a self-made cylindrical steel tube furnace (SiC heater: length 1000 mm, diameter 100 mm) under N_2 gas (99.999%) and kept at a target temperature for 1 h to obtain 2 g of carbonized polymer-based HC.

2.2. Steam Activation Optimizaiton

Activation of the HC was performed using steam as the physical reagent. Steam activation was performed in the same steel tube used for carbonization. The HC was heated (10 °C/min) to the required activation temperature (900 or 1000 °C) under N_2 (300 mL/min), before the N_2 was replaced with H_2O (0.5 mL/min, liquid) with various activation times. The sample was then allowed to cool under N_2 (300 mL/min). The samples were named as activated polymer-based hard carbons (APHS) and named APHS-activation temperature-activation time sequence according to activation conditions. The prepared AC was pulverized to 10 μm size using a ball mill (Pulverisette 23, Idar-Oberstein, Germany).

2.3. Characterization

The N_2 adsorption isotherms of APHS were measured with a BELSORP-max (BEL JAPAN, Toyonaka, Japan) at liquid nitrogen temperature. For pore analysis, all samples were degassed at 573 K for 6 h with the residual pressure maintained at 10^{-3} torr, or less. The specific surface area was calculated for the relative pressure interval of 0.03–0.19 using the Brunauer-Emmett-Teller (BET) equation. [28] The total pore volume, V_{Total}, was calculated from N_2 adsorption data as the volume of liquid N_2 at a relative pressure of 0.99. The mesopore volume, V_{Meso}, was determined by the Barrett-Joyner-Halenda (BJH) method, and the micropore volume, V_{Micro}, was obtained by subtraction of the mesopore volume from the total pore volume. Micropore and mesopore size and distribution were calculated using the non-local density functional theory (NLDFT) [29] and BJH method [30], respectively. The microstructure of the APHS was determined using an X-ray diffractometer (XRD, X'Pert Pro Diffractometer, PANalytical, Almelo, The Netherlands), employing a Rigaku SmartLab X-ray diffractor with a customized auto-mount and a Cu Kα (λ = 1.5406 Å) radiation source. Diffraction patterns were collected within the diffraction angles from 5° to 90° at a rate of 2°/min. The interlayer spacing (d_{002}, d_{10l}) of the samples were calculated using Bragg's Law (2dsinθ = nλ) to the position of the (002) and (10*l*) peak, respectively. The size of crystalline under the Scherrer equation [31] can be expressed as follows:

$$L = \frac{K\lambda}{B\cos\theta} \tag{1}$$

In this equation, the constant K is 0.9 and 1.84 when calculated from crystalline height (L_c) and crystalline diameter (L_a), respectively. λ is the wavelength of the X-ray, and B is the full width at half maximum (FWHM) calculated from the radian.

2.4. Electrochemical Measurements

The slurry was prepared by mixing 80 wt% active materials (activated HC, APHS), 10 wt% conductive agents (carbon black, Super P, TIMCAL, Bodio, Switzerland), 10 wt% binder (carboxymethyl cellulose and styrene-butadiene rubber) dispersed in water. The slurries were coated on the aluminum foil using a doctor blade technique. The thickness of the coating layer was controlled to 127 μm. The coated foils were then first dried in an oven at 80 °C for over one hour. To remove any remaining water, the electrodes were further dried in a vacuum oven at 120 °C overnight. After the drying was completed, the electrodes were roll pressed to a thickness of less than 80 μm using a two-roll press at 80 °C.

For electrochemical testing, a CR2032 coin cell consisting of two punched electrodes 12 mm in diameter, punched cellulose paper separators 16 mm in diameter (NKK, Kochi, Japan), and 1M $(C_2H_5)_4NBF_4$/propylene carbonate (1M $TEABF_4$/PC) were used. All cell assembly was carried out in a dry room where the dew point was below 60.0 °C.

The fifth cycle of galvanostatic charge/discharge (GCD) and the second cycle of cyclic voltammetry (CV) were used to evaluate the electrochemical performance of the samples. The GCD was charged at 10 mA and then discharged at 2 mA (0.1–2.4 V) for the coin cell using a charge/discharge tester (Maccor 4300 K, Maccor, Tulsa, OK, USA). CV and impedance spectrometry (EIS) were analyzed using a potentiostat (Bio-Logic VSP, Bio-Logic Science Instruments, Seyssinet-Pariset, France). CV studies were performed in the same potential range of GCD at a scan rate of 30 mV/s. The impedance Nyquist plots were recorded in the frequency range 10 mHz to 300 kHz. The cells produced were measured based on the capacitance per unit weight, and calculated only using the weight of active materials (F/g). The specific capacitance was calculated according to the GCD based on the following equation:

$$C_g = i\Delta t/m\Delta V \tag{2}$$

where i is the discharge current (A), Δt is the discharge time (s), m is the mass of the electrode (g), and ΔV is the potential difference (V).

3. Results

3.1. X-ray Diffraction Analysis

It is well known that XRD is a powerful technique for revealing detailed and precise microstructure, such as the interlayer spacing (d_{002}) of AC, which is composed of thin layers with the same atomic positions as graphite within the layers. Figure 1 shows the X-ray diffraction profiles of the APHS prepared under different activation conditions. The APHS exhibits very broad diffraction peaks and the absence of a sharp peak reveals a predominant HC structure. The XRD pattern of the AC was very similar to that of the HC. In this work, the interlayer spacing of the APHS is about 3.62 to 3.76 Å, and that of the HC is 3.68 Å, as seen in Figure 2.

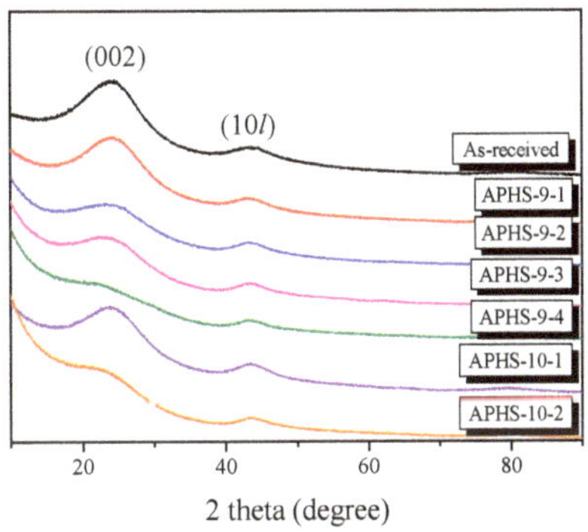

Figure 1. X-ray diffraction patterns of activated polymer-based hard carbon under various steam activation conditions.

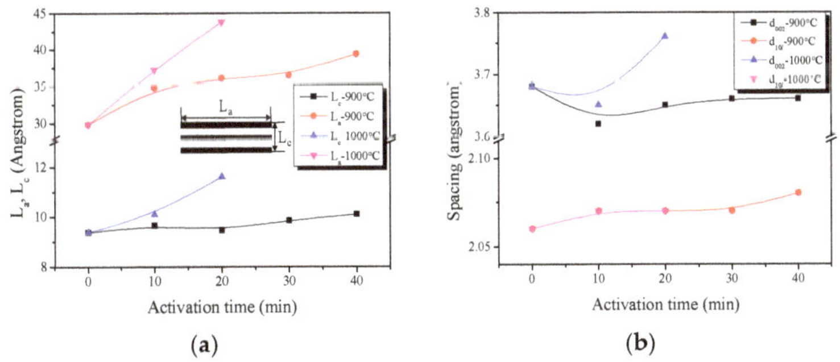

Figure 2. Structural characteristics of the activated polymer-based hard carbon as a function of various steam activation conditions: (**a**) structural parameters; (**b**) interplanar distance.

The oxidation behaviors of HC tend to operate in the amorphous regions and graphite edges [11]. It is widely known that amorphous carbon can be more easily oxidized than crystalline carbon can during the activation reactions [11–13]. If carbon atoms in the amorphous region of HC were removed first, the overall crystallinity of the HC could be increased due to decrease in a portion of the amorphous region, resulting in the increase of L_c or L_a in the final APHS.

As shown in Figure 2, L_c does not exhibit a large change at an activation temperature of 900 °C, but does gradually increase with the activation time. However, L_a increases steadily with increasing activation time. In particular, the largest increase is found for the sample that was activated for 10 min, followed by 30 to 40 min. It is believed that much oxidation of amorphous parts or small crystallites occurred in this section. This was confirmed in the yield of Table 1. Both L_c and L_a increase significantly at the activation temperature of 1000 °C. These results suggest that the activation reaction at high temperature is strong and rapid.

Table 1. Textural properties of polymer-based hard carbon activated under various steam activation conditions.

Sample	Activation Conditions	Yield [2] (%)	S_{BET} [3] (m^2/g)	V_{Total} [4] (cm^3/g)	V_{Meso} [5] (cm^3/g)	V_{Micro} [6] (cm^3/g)	C_g [7] (F/g)
As-received	-	100	50	0.03	0.01	0.02	0.8
APHS-9-1 [1]	900 °C × 10 min	60	1170	0.48	0.02	0.46	0.8
APHS-9-2	900 °C × 20 min	34	1900	0.84	0.16	0.68	73.3
APHS-9-3	900 °C × 30 min	33	2030	0.91	0.17	0.74	115.8
APHS-9-4	900 °C × 40 min	17	2240	1.16	0.38	0.78	136.2
APHS-10-1	1000 °C × 10 min	48	1240	0.53	0.06	0.47	26.4
APHS-10-2	1000 °C × 20 min	15	2410	1.22	0.47	0.75	108.2
YP-50F	Steam activation	-	1720	0.79	0.15	0.64	91.8
MSP-20	KOH activation	-	2260	0.97	0.11	0.86	125

[1] APHS: polymer-based hard carbon activated using steam activation. [2] Yield: $\frac{\textit{Weight of activated sample}}{\textit{Weight of hard carbon inputted}} \times 100$. [3] S_{BET}: Specific surface area by the Brunauer-Emmett-Teller (BET) method. [4] V_{Total}: Total pore volume; BET method. [5] V_{Meso}: Mesopore volume by the Barrett-Joyner-Halenda (BJH) method. [6] V_{Micro}: Micropore volume; $V_{Total} - V_{Meso}$. [7] C_g: Specific capacitance.

The interlayer spacing exhibited a large change at d_{002} and a small change at d_{10l}. The interlayer spacing decreased up to APHS-9-1, and then increased. The increase in the interplanar distance d_{002} was clearly observed with increase in the activation time. More specifically, the distance decreased from 3.68 Å for the HC to 3.62 Å for APHS-9-1. The amorphous oxidation within HC is presumed to cause this decrease in the interplanar distance of d_{002}. In addition, a similar trend is observed in APHS-10-1. After amorphous oxidation, the small crystallite is oxidized and increases the interplanar distance of d_{002}. According to the activation progression, the specific surface area of APHS was increased by oxidation of small crystallites. Based on these observations, it can be postulated that long activation time and high temperature burn off the less ordered components and small crystallites, resulting in the increased specific surface area of the APHS.

3.2. Adsorption Isotherm and Textural Properties

The adsorption isotherms of N_2 at 77 K contain a large amount of information related to pore-structure. Because changes in the isotherms imply an alteration of pore structures; their measurements are the basis upon which to estimate the pore structure parameters. The specific surface area and pore structures of the APHS before and after activation were determined using N_2 adsorption/desorption isotherms (Figure 3). There are six kinds of adsorption isotherm patterns according to the International Union of Pure and Applied Chemistry (IUPAC) classification [32], and each is indicated by a distinct pore structure. According to the IUPAC standards, the curves in Figure 3 all have Type I shape, largely consisting of micropores; longer activation time was associated with increased specific surface area and quantity of mesopores.

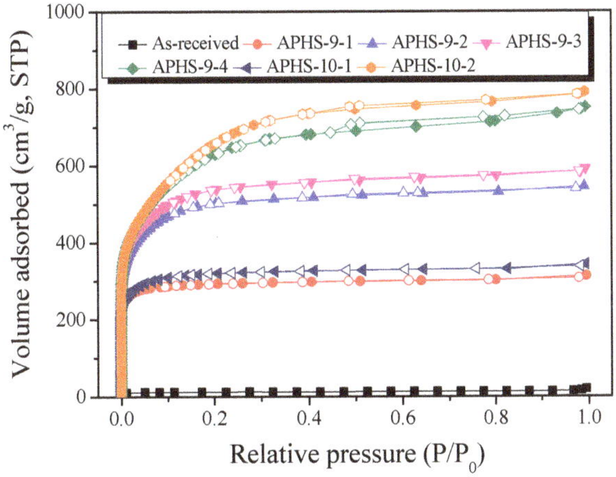

Figure 3. N_2/77 K adsorption-desorption isotherm curves of activated polymer-based hard carbon under various steam activation conditions.

When the hysteresis curves were examined, hysteresis curves were rarely observed except for APHS-9-4 and APHS-10-2. This seems to be because the mesopores of these two samples were well-developed, so that many pores underwent interior development so that the shape of such pores changed to the shape of a jar.

The effect of the steam activation conditions on the specific surface area, total pore volume, and micropore volume are given in Table 1. The specific surface area and total pore volume increase with increasing activation temperature and time. As activation time and temperature increase during the steam activation process, the specific surface area increases from 50 to 2410 m^2/g, and the total pore volume increases from 0.03 to 1.22 cm^3/g.

When the pore characteristics were examined in detail at the activation temperature of 900 °C, APHS activated for 10 min developed only micropores. At up to 40 min of activation, both micropores and mesopores increased, with the former increasing from 0.02 to 0.78 cm^3/g and the latter increasing from 0.01 to 0.38 cm^3/g. In the case of APHS activated for 40 min, the volume of micropores increased from 0.74 to 0.78 cm^3/g, while that of mesopores increased from 0.17 to 0.38 cm^3/g. Because the activation yield was greatly reduced, mainly mesopores were increased. This is because micropores collapsed and developed into mesopores due to the activation reaction induced by steam oxidation. However, as the volume of micropores continued to increase, it was recognized that new micropores continued to be formed as the activation time increased. Activation at 1000 °C was observed to develop pores more quickly than activation at 900 °C. APHS-10-2 sample exhibited the highest specific surface area and total pore volume.

Figure 4 exhibits the mesopore size distributions after application of the BJH method. All samples have their highest intensity peak around 2.5 nm pore diameter. APHS-9-1 and APHS-10-1, and APHS-9-4 and APHS-10-2, were produced using different activation processes, but have similar specific surface areas. However, the APHS produced at higher temperatures exhibits greater mesopore volume. That is, the high activation temperature produced APHS that was richer in mesopores.

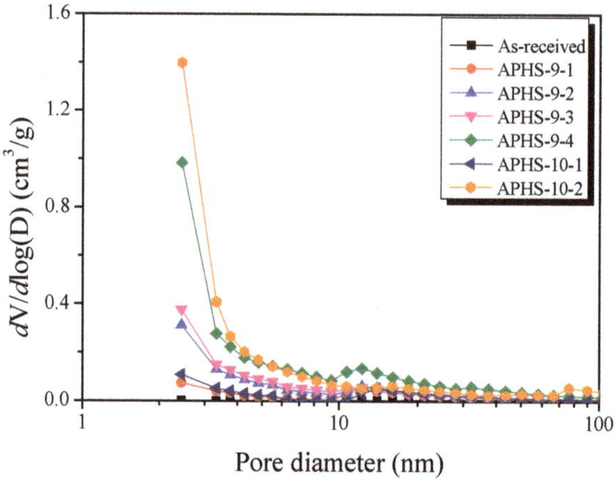

Figure 4. Pore size distribution of polymer-based hard carbon activated under various steam activation conditions using the BJH method.

Figure 5 exhibits the micropore size distributions created by the NLDFT method. As activation time increases, narrow micropores develop. High micropore volume is observed from up to 20 min of activation time in the diameter range from 0.75 to 1.0 nm. When activation time exceeds 20 min and stretches to 30 min, micropore volume decreases in the diameter range from 0.75 to 1.0 nm, while the micropore volume increases in the diameter range from 1.0 nm and more. Moreover, the pore size distribution becomes broader.

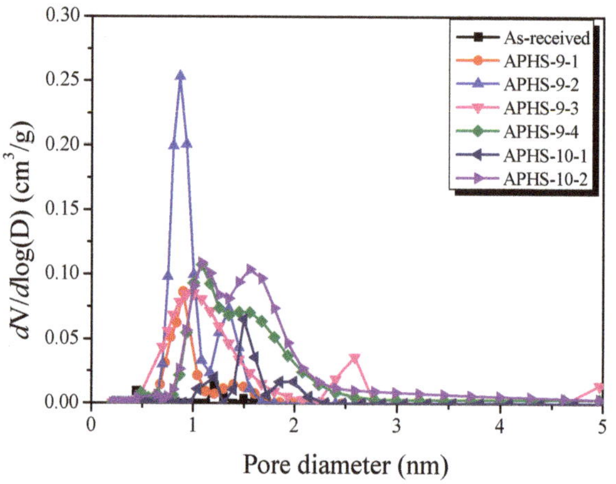

Figure 5. Pore size distribution of polymer-based hard carbon activated under various steam activation conditions using the non-local density functional theory method.

It is widely known that steam activation is a process that develops pores by oxidizing the carbon atoms of the precursors [13,16]. During the initial increase of activation time, micropores with small pore diameters are formed first as a result of the reaction between steam and carbon. With increasing activation time, the oxidation of crystallites increases, leading to increased specific surface areas. However, with additional activation time, the pores developed in the initial state start to deepen, enlarge, and perhaps merge, resulting in the observed increase in the average pore width. In addition, new micropores develop within the mesopores, and the specific surface area continues to increase.

3.3. Electrochemical Characterization

The electrochemical properties, including the galvanostatic discharge curves of the APHS electrodes, were studied using an electrolyte of 1M $TEABF_4$/PC. Theoretically, the capacitance of the AC is proportional to its specific surface area [6]. However, only the surface of the pores that the ions can access can contribute to double layer capacitance [5,7]. In particular, the pore size distribution has been considered the most important parameter, because the accessibility of ion molecules in an electrolyte strongly depends on the pore size of the electrodes. The sizes of non-solvated ions and the sizes of solvated ions in 1M $TEABF_4$/PC is 0.34 to 1.40 nm [7]. Therefore, mesopores are more useful than micropores for EDLCs, especially for non-aqueous and ionic liquid (IL) EDLCs with larger ions. Baek et al., reported a close relationship between specific capacitance and the pore size of activated carbon (2–5 nm and >5 nm) in 1M $TEABF_4$/PC [7].

Figure 6a exhibits a change in the specific capacitance according to the charge/discharge cycle. All APHS exhibit stable initial specific capacitance. The specific capacitance is estimated from the galvanostatic charge/discharge curve (Figure 6b), which corresponds to the calculated specific capacitance of the electrode. The specific capacitance of all the APHS samples increases with increasing activation times and temperature. APHS-9-1 has very low capacity due to small pore size and low pore volume. The mobility of the ions within the pores is greatly influenced by pore size. If the pores are too small to allow easy access to electrolyte ions, they will not contribute to double-layer capacitance. APHS-9-2 has specific capacitance higher than that of APHS-9-1 because micropores and mesopores were significantly increased by activation. APHS-9-3 exhibits micropore and mesopore volume slightly increased over those of APHS-9-2; however, the specific capacitance increases significantly from 73.2 to 115.8 F/g. As discussed in Figure 5, the main pore diameter is distinctively enlarged due to expansion of the previously generated micropores and to collapse of micropore walls. Therefore, it is considered that the adsorption capacity of ions would be increased to provide easy access to electrolyte ions by the enlarged pores. In this work, APHS-9-4 exhibits the best mesopore fraction and 136.1 F/g of energy storage ability. The galvanostatic discharge curve of APHS-9-4 exhibited a straight line, typical of EDLC, with a non-IR drop. (Because the discharge data was not converted to the value per weight of the electrode material, they cannot be compared with the absolute value.) This is about 148% and 108% higher than the YP50F (95 F/g, coconut shell origin, physical activation with steam, Kuraray Chemical Co., LTD., Osaka, Amagasaki, Japan) and MSP20 (125 F/g, phenol resin origin, chemical activation with KOH, Kansai Coke and Chemicals, Japan) electrode capacity, respectively [16,33]. The porosity of APHS-9-1 and APHS-10-1 are very similar, but the specific capacitance of APHS-10-1 is higher than that of APHS-9-1. In Figure 5, the pore diameter of APHS-10-1 is larger than that of APHS-9-1, with easy access to electrolyte ions. APHS-10-2 has the highest specific surface area and mesopore volume. However, APHS-10-2 exhibits lower specific capacitance than APHS-9-3 and APHS-9-4. As seen in Figure 5, the pore width of APHS-10-2 is wider than that of APHS-9-3 and APHS-9-4. Therefore, APHS-10-2 has low capacitance despite the fact that APHS-10-2 has the best pore properties.

Figure 7 exhibits the result of plotting the pore volume according to pore diameter in 0.5 nm units using NLDFT, and then plotting the coefficient of determination with specific capacitance. An empirical linear fit was used to evaluate the contributions from each divided pore volume. The *X*-axis is exhibited using the average value of each pore size distribution. The R^2 coefficient of determination ($R^2 = 1 - SS_{res}/SS_{tot}$, SS_{res} is residual sum of squares and SS_{tot} is total sum of squares) exhibits a trend of increase and then decrease again after the occurrence of the highest value in the results (1.5–2.5 nm pore diameter).

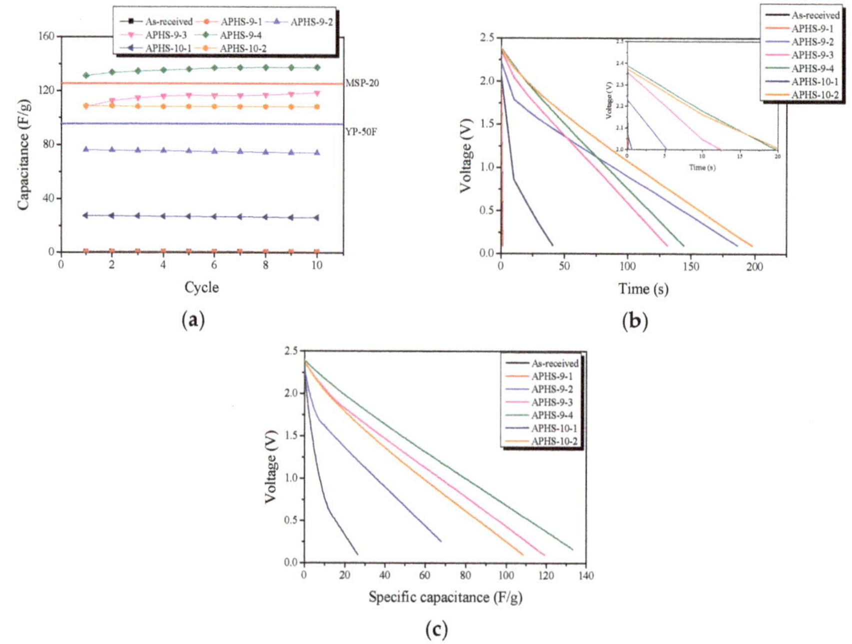

Figure 6. Electrochemical performance of the activated polymer-based hard carbon as a function of a cycle under various steam activation conditions: (**a**) initial specific capacitance; (**b**) discharge curves; (**c**) specific capacitance curves. Because the discharge data was not converted to the value per weight of electrode material, they cannot be compared with absolute value.

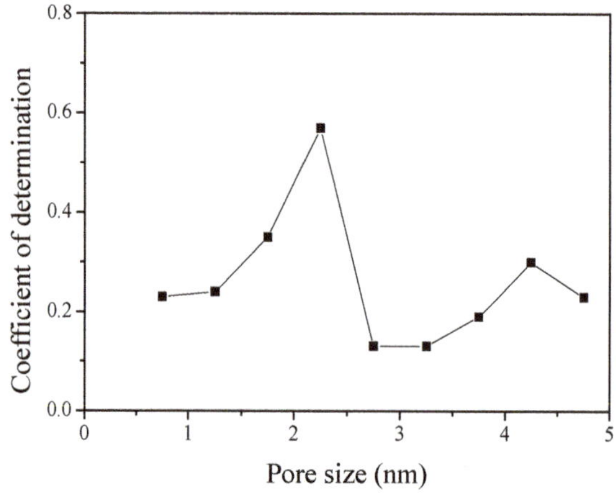

Figure 7. Correlation between the specific capacitance of activated polymer-based hard carbon with various pore volume. The *X*-axis exhibits the average pore size distribution. It was plotted using the pore volume according to the pore diameter in 0.5 nm units and the average value of each pore size distribution.

The typical cyclic voltammograms (CV) of the capacitor cells are shown in Figure 8. All the CV curves are rectangular without obvious redox current on both positive and negative sweeps over the whole potential range of 0.1 to 2.5 V at 30 mV/s in the organic electrolyte (1M $TEABF_4$/PC). At longer activation time, CV curves gradually change into rectangles. Moreover, APHS-9-3 and APHS-9-4 show a symmetric, quasi-rectangular shape profile typical of ideal EDLCs. This is consistent with the results of pore characteristic analysis and seems to be because the pores are better developed than those of other ACs. Generally, the specific capacitance of an electrode is in proportion to the integrated area of its CV profile under the same scan rate and voltage window (i.e., the larger the integrated area is, the greater the specific capacitance). As activation time increases, the area of the CVs tends to increase. The capacitance of the electrodes calculated from the CV curves decreases in the following order: APHS-9-4 > APHS-9-3 > APHS-10-2 > APHS-10-1 > APHS-9-2 > APHS-9-1. Among the carbon samples, APHS-9-4 shows the highest specific capacitance due to its uniform mesopores and high specific surface area.

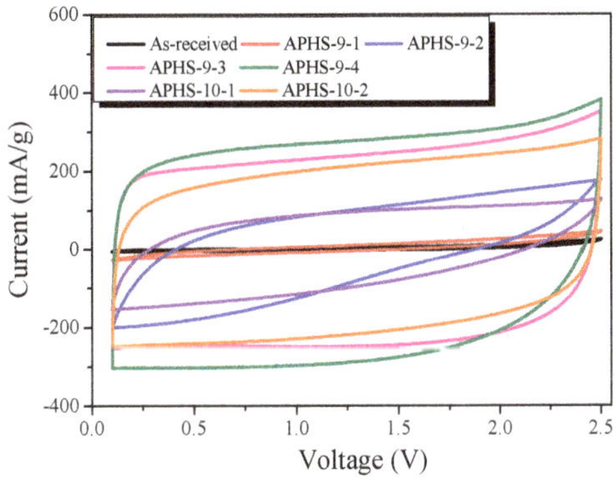

Figure 8. Cyclic-voltammetry curves of polymer-based hard carbon activated under various steam activation conditions.

Impedance spectroscopy, which distinguishes the resistance and capacitance of devices, was used to perform a comprehensive analysis of the EDLC cells. Figure 9 shows corresponding impedance plots for an organic (1M $TEABF_4$/PC) electrolyte. The frequency was swept from 10 mHz to 300 kHz. Each curve presents a depressed semicircle in the middle (high frequency region) and a nearly perpendicular line in a low-frequency region. In general, the impedance spectrum of EDLCs consists of three frequency-sensitive regions showing the characteristic shape of a $Z'' = f(Z')$ curve. The semicircle present at high frequencies is due to (i) electrode porosity and (ii) the charge transfer resistance of possible pseudo-capacity contributing to the total observed capacity. The electrolyte resistance is in series with the latter resistance. The electrolyte resistance influences not only the shape of the plot but also the value of ohmic resistance at which the vertical line cuts off the real resistance axis. The middle-frequency region, represented by the 45° line, is rather due to the frequency dependent resistance $R(\omega)$ associated with electrolyte penetration of the electrode pores.

All samples had the same sized semicircles at 900 °C activation temperature. Additionally, as the activation temperature increased, the size of the semicircle decreased. The higher the activation temperature, the larger were the diameter pores that developed. The charge transfer resistance decreased due to the easy movement of ions.

Pore shapes affect impedance behavior, specifically, the form of the impedance [34]. At 900 °C activation temperature, the morphology of the pore changes from a cylindrical shape (Type I) to a jar shape (Type IV) as the activation time increases. Moreover, the pore shape changes from wedge shape (Type V) to jar shape with increasing activation time at 1000 °C. At 900 °C, activation starts with pores of cylindrical form because of the slow oxidation of the graphite crystals, but at 1000 °C activation, because the oxidation of graphite crystals occurs rapidly, they develop from wedge-shaped pores. In the case of physical activation, small crystals inside the activated carbon are oxidized to form pores. Therefore, the pores of activated carbon are narrow at the entrance, and many additional pores develop inside. Finally, the pore shape of the activated carbon becomes a jar shape.

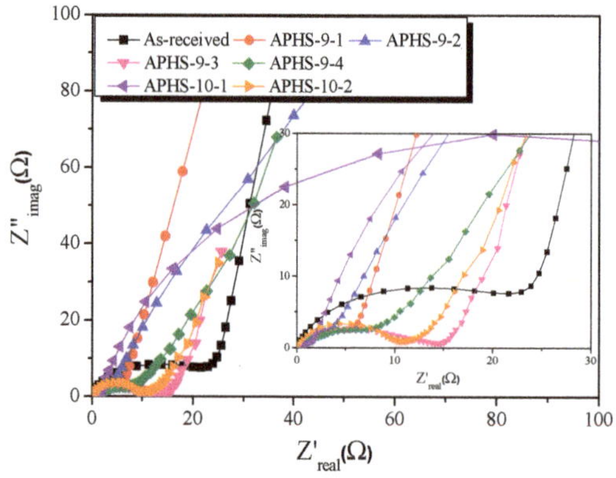

Figure 9. Nyquist plots of polymer-based hard carbon activated under various steam activation conditions.

As the activation time increases, the pores develop toward the inside of the activated carbon. Therefore, as the activation time increases, the ions must move deep within the activated carbon for adsorption. As the electrolyte ion movement time increases, resistance is generated, and thereby, the slope of the straight line decreases.

4. Conclusions

In this work, a polymeric precursor was modified using variations of steam activation methods to prepare APHSs with high specific surface area and a mesopore-rich structure. The specific surface area and specific capacitance of the physical APHS samples was up to 2240 m^2/g and 136 F/g, respectively.

The activation condition (time and temperature) affected the pore structure of APHS. During the initial increase of the activation time, micropores with narrow PSD were formed by oxidation of amorphous areas or small crystallites. As the activation time increased, the oxidation of crystallites increased, leading to increased specific surface areas and mesopore volume. As the activation time increased further, the oxidation of crystallites increased, leading to increased micropore and mesopore volume. With additional extension of the activation time, the pores that initially developed started to deepen, enlarge, and perhaps merge, resulting in the observed increase in the average pore width. When the activation temperature increased from 900 to 1000 °C, the oxidation rate of APHS was found to increase about twice. As a result, at the activation temperature of 1000 °C, APHS exhibited broad PSD curves and wide pore diameter.

The specific capacitance was significantly dependent on the pore size distribution according to activation conditions. It was confirmed that the correlation between the specific capacitance in 1M TEABF4/PC and the pore characteristics of the APHS was determined by pores of diameter 1.5–2.5 nm. The specific capacitance of APHS-9-4 was higher than that of APHS-10-2, which had the highest specific surface area and mesopore volume. These results suggest that the pore structure of APHS-9-4 is better optimized than is the pore structure of APHS-10-2 in 1M $TEABF_4$/PC. In conclusion, APHS, created using the physical activation method, exhibited better specific capacitance than did MSP20 created using the chemical activation method.

Author Contributions: Conceptualization, S.-J.P. and B.-J.K.; methodology, S.-J.P. and B.-J.K.; validation, K.-H.A., S.-J.P. and B.-J.K.; formal analysis, H.-M.L. and B.-J.K.; investigation, K.-H.A.; resources, K.-H.A. and B.-J.K.; data curation, H.-M.L.; writing—original draft preparation, H.-M.L.; writing—review and editing, B.-J.K.; visualization, B.-J.K.; supervision, B.-J.K.; project administration, B.-J.K.; funding acquisition, B.-J.K.

Funding: This research was supported by the Nano·Material Technology Development Program through the National Research Foundation of Korea (NRF) funded by the Ministry of Science and ICT (Project no. 2018M3A7B9086636). This research was financially supported by the Ministry of SMEs and Startups (MSS), Korea, under the "Regional Specialized Industry Development Program (P0003187)" supervised by the Korea Institute for Advancement of Technology (KIAT).

Conflicts of Interest: The authors declare no conflict of interest.

References

1. Miller, J.R.; Simon, P. Electrochemical capacitors for energy management. *Science* **2008**, *321*, 651–652. [CrossRef]
2. Pham, D.T.; Lee, T.H.; Luong, D.H.; Yao, F.; Ghosh, A.; Le, V.T.; Kim, T.H.; Li, B.; Chang, J.; Lee, Y.H. Carbon nanotube-bridged graphene 3D building blocks for ultrafast compact supercapacitors. *ACS Nano* **2015**, *9*, 2018–2027. [CrossRef]
3. Sharma, P.; Bhatti, T.S. A review on electrochemical double-layer capacitors. *Energy Convers. Manag.* **2010**, *51*, 2901–2912. [CrossRef]
4. Chmiola, J.; Yushin, G.; Gogotsi, Y.; Portet, C.; Simon, P.; Taberna, P.L. Anomalous increase in carbon capacitance at pore sizes less than 1 nanometer. *Science* **2006**, *313*, 1760–1763. [CrossRef] [PubMed]
5. Xu, B.; Wu, F.; Chen, R.; Cao, G.; Chen, S.; Zhou, Z.; Yang, Y. Highly mesoporous and high surface area carbon: A high capacitance electrode material for EDLCs with various electrolytes. *Electrochem. Commun.* **2008**, *10*, 795–797. [CrossRef]
6. Barbieri, O.; Hahn, M.; Herzog, A.; Kotz, R. Capacitance limits of high surface area activated carbons for double layer capacitors. *Carbon* **2005**, *43*, 1303–1310. [CrossRef]
7. Baek, J.; An, K.H.; Chung, D.C.; Kim, B.J. Correlation studies between pore structure and electrochemical performance of activated polymer-based hard carbon with various organic and aqueous electrolytes. *Int. J. Energy Res.* **2018**, *42*, 2927–2939. [CrossRef]
8. Jung, J.Y.; Lee, Y.S. Electrochemical properties of KOH-activated lyocell-based carbon fibers for EDLCs. *Carbon Lett.* **2018**, *27*, 112–116.
9. Xu, Y.; Chen, X.; Wu, D.; Luo, Y.; Liu, X.; Qian, Q.; Xiao, L.; Chen, Q. Carbon molecular sieves from soybean straw-based activated carbon for CO_2/CH_4 separation. *Carbon Lett.* **2018**, *25*, 68–77.
10. Lee, H.W.; Kim, Y.M.; Kim, S.; Ryu, C.; Park, S.H.; Park, Y.K. Review of the use of activated biochar for energy and environmental applications. *Carbon Lett.* **2018**, *26*, 1–10.
11. Baek, J.; Lee, H.M.; Roh, J.S.; Lee, H.S.; Kang, H.S.; Kim, B.J. Studies on preparation and applications of polymeric precursor-based activated hard carbons: I. Activation mechanism and microstructure analyses. *Microporous Mesoporous Mater.* **2016**, *219*, 258–264. [CrossRef]
12. Baek, J.; Shin, H.S.; Chung, D.C.; Kim, B.J. Studies on the correlation between nanostructure and pore development of polymeric precursor-based activated hard carbons: II. Transmission electron microscopy and Raman spectroscopy studies. *J. Ind. Eng. Chem.* **2017**, *54*, 324–331. [CrossRef]
13. Lee, H.M.; Heo, Y.J.; An, K.H.; Jung, S.C.; Chung, D.C.; Park, S.J.; Kim, B.J. A study on optimal pore range for high pressure hydrogen storage behaviors by porous hard carbon materials prepared from a polymeric precursor. *Int. J. Hydrog. Energy* **2018**, *43*, 5894–5902. [CrossRef]
14. Lee, H.M.; Kang, H.R.; An, K.H.; Kim, H.G.; Kim, B.J. Comparative studies of porous carbon nanofibers by various activation methods. *Carbon Lett.* **2013**, *14*, 180–185. [CrossRef]
15. Lee, H.M.; An, K.H.; Kim, B.J. Effects of carbonization temperature on pore development in polyacrylonitrile-based activated carbon nanofibers. *Carbon Lett.* **2014**, *15*, 146–150. [CrossRef]
16. Lee, H.M.; Chung, D.C.; Jung, S.C.; An, K.H.; Park, S.J.; Kim, B.J. A study on pore development mechanism of activated carbons from polymeric precursor: Effects of carbonization temperature and nano crystallite formation. *Chem. Eng. J.* **2019**, in press. [CrossRef]
17. Wang, J.; Kaskel, S. KOH activation of carbon-based materials for energy storage. *J. Mater. Chem.* **2012**, *22*, 23710–23725. [CrossRef]
18. Kan, Y.; Yue, Q.; Liu, S.; Gao, B. Effects of Cu and CuO on the preparation of activated carbon from waste circuit boards by H_3PO_4 activation. *Chem. Eng. J.* **2018**, *331*, 93–101. [CrossRef]

19. Acharya, J.; Sahu, J.N.; Sahoo, B.K.; Mohanty, C.R.; Meikap, B.C. Removal of chromium(VI) from wastewater by activated carbon developed from tamarind wood activated with zinc chloride. *Chem. Eng. J.* **2009**, *150*, 25–39. [CrossRef]
20. Lee, H.M.; Kim, H.G.; An, K.H.; Kim, B.J. The effect of CO_2 activation on the electrochemical performance of coke-based activated carbons for supercapacitors. *J. Nanosci. Nanotechnol.* **2015**, *15*, 8797–8802. [CrossRef]
21. Shrestha, M.; Amatya, I.; Wang, K.; Zheng, B.; Gu, Z.; Fan, Q.H. Electrophoretic deposition of activated carbon YP-50 with ethyl cellulose binders for supercapacitor electrodes. *J. Energy Storage* **2017**, *13*, 206–210. [CrossRef]
22. Pode, R. Potential applications of rice husk ash waste from rice husk biomass power plant. *Renew. Sustain. Energy Rev.* **2016**, *53*, 1468–1485. [CrossRef]
23. Wu, J.; Hong, I.; Park, S.M.; Lee, S.Y.; Kim, M.S. Electrochemical properties of EDLC electrodes prepared by acid and heat treatment of commercial activated carbons. *Carbon Lett.* **2008**, *9*, 137–144. [CrossRef]
24. Kim, J.A.; Park, I.S.; Seo, J.H.; Lee, J.J. A development of high power activated carbon using the KOH activation of soft carbon series cokes. *Trans. Electr. Electron. Mater.* **2014**, *15*, 81–86. [CrossRef]
25. Hyman, J.B.; Sterr, W.R. Catalin Corp of America. Low Ash Phenol-Formaldehyde Resins and Process of Preparation. U.S. Patent 2865875, 23 December 1958.
26. Wu, F.C.; Tseng, R.L.; Hu, C.C. Comparisons of pore properties and adsorption performance of KOH-activated and steam-activated carbons. *Microporous Mesoporous Mater.* **2005**, *80*, 95–106. [CrossRef]
27. Lee, J.H.; Choi, J.W.; Kim, S.A.; Lee, J.M.; Lee, J.M.; Kim, J.G. Negative Active Material for Lithium Secondary Battery, Preparing Method Thereof and Lithium Secondary Battery Using the Same. KR Patent 101375688, 12 March 2014. [CrossRef]
28. Brunauer, S.; Emmett, P.H.; Teller, E. Adsorption of gases in multi molecular layers. *J. Am. Chem. Soc.* **1938**, *60*, 309–319. [CrossRef]
29. Kierlik, E.; Rosinberg, M.L. Free-energy density functional for the in homogeneous hard-sphere fluid: Application to interfacial adsorption. *Phys. Rev. A* **1990**, *42*, 3382–3387. [CrossRef] [PubMed]
30. Barrett, E.P.; Joyner, L.G.; Halenda, P.P. The determination of pore volume and area distributions in porous substances. I. Computations from nitrogen isotherms. *J. Am. Chem. Soc.* **1951**, *73*, 373–380. [CrossRef]
31. Biscoe, J.; Warren, E. An X-ray study of carbon black. *J. Appl. Phys.* **1942**, *13*, 364–371. [CrossRef]
32. Sing, K.S.W.; Everett, D.H.; Haul, R.A.W.; Moscou, L.; Pierotto, R.A.; Rouquerol, J. Reporting physisorption data for gas/solid systems with special reference to the determination of surface area and porosity. *Pure Appl. Chem.* **1985**, *57*, 603–619. [CrossRef]
33. Jung, M.J.; Jeong, E.; Cho, S.; Yeo, S.Y.; Lee, Y.S. Effects of surface chemical properties of activated carbon modified by amino-fluorination for electric double-layer capacitor. *J. Colloid Interface Sci.* **2012**, *381*, 152–157. [CrossRef]
34. Kaiser, H.; Beccu, K.D.; Gujahr, M.A. Abschätzung der porenstruktur poröser elektroden aus impedanzmessungen. *Electrochim. Acta* **1976**, *21*, 539–543. [CrossRef]

© 2019 by the authors. Licensee MDPI, Basel, Switzerland. This article is an open access article distributed under the terms and conditions of the Creative Commons Attribution (CC BY) license (http://creativecommons.org/licenses/by/4.0/).

nanomaterials

MDPI

Article

Large Scale Process for Low Crystalline MoO_3-Carbon Composite Microspheres Prepared by One-Step Spray Pyrolysis for Anodes in Lithium-Ion Batteries

Jung Sang Cho

Department of Engineering Chemistry, Chungbuk National University, Chungbuk 361-763, Korea; jscho@cbnu.ac.kr; Tel.: +82-43-261-2489; Fax: +82-43-262-2380

Received: 22 February 2019; Accepted: 27 March 2019; Published: 3 April 2019

Abstract: This paper introduces a large-scale and facile method for synthesizing low crystalline MoO_3/carbon composite microspheres, in which MoO_3 nanocrystals are distributed homogeneously in the amorphous carbon matrix, directly by a one-step spray pyrolysis. The MoO_3/carbon composite microspheres with mean diameters of 0.7 μm were directly formed from one droplet by a series of drying, decomposition, and crystalizing inside the hot-wall reactor within six seconds. The MoO_3/carbon composite microspheres had high specific discharge capacities of 811 mA h g^{-1} after 100 cycles, even at a high current density of 1.0 A g^{-1} when applied as anode materials for lithium-ion batteries. The MoO_3/carbon composite microspheres had final discharge capacities of 999, 875, 716, and 467 mA h g^{-1} at current densities of 0.5, 1.5, 3.0, and 5.0 A g^{-1}, respectively. MoO_3/carbon composite microspheres provide better Li-ion storage than do bare MoO_3 powders because of their high structural stability and electrical conductivity.

Keywords: molybdenum oxide; carbon composite; spray pyrolysis; anodes; lithium-ion batteries

1. Introduction

Lithium-ion batteries (LIBs) have been attractive as the most important type of power source for energy-storage system, electric vehicles, and other electronic devices because of their high specific capacities and energy densities [1–3]. Transition-metal oxides (TMOs) with high theoretical energy capacities have been widely applied as replacement anodes for the current graphite of LIBs [4–6]. However, the low intrinsic electric conductivity and the large volume expansion of TMOs during a charge/discharge cycle result in rapid capacity fading, which hinders the commercial application of TMOs for anodes in current LIBs [7,8]. To solve these problems, compositing TMOs with carbonaceous materials has been regarded as a possible solution. Carbon could effectively buffer the stress induced by the large volume change of TMOs during the fast charging–discharging process and improve the electrical conductivity of the anodes [9–11]. Additionally, a carbon matrix could prevent the aggregation of the active materials during repeated cycles by surrounding them, which increases the structural stability of anode materials [12,13]. Therefore, various synthesis strategies for TMOs/carbon composites have been introduced [14–18]. Cho et al. [14] prepared multiroom-structured metal–carbon hybrid microspheres containing empty voids of several tens of nanometers by liquid–liquid phase segregation because of the incongruent melting of the metal salt and dextrin during the spray pyrolysis. The discharge capacity of the multiroom-structured Co_3O_4–C hybrid microspheres for LIBs at a current density of 3 A g^{-1} for the 150th cycle was 1243 mA h g^{-1}. Zhang et al. [15] also prepared TiO_2–graphene composite nanofibers by a simple electrospinning process. The cell assembled with TiO_2–graphene composite nanofibers as an anode retained 84% of the reversible capacity after 300 cycles at a current density of 150 mA g^{-1}, which is 25% higher than bare TiO_2 nanofibers did under the same test conditions. Bhaskar et al. [16] prepared MoO_2/multiwalled carbon nanotubes (MWCNTs) composed of spherical flowerlike nanostructures of MoO_2, interconnected by MWCNTs by a one-step hydrothermal route. The one-dimensional electron-transport pathways provided by MWCNTs, which are in direct contact with the MoO_2 nanostructures, imparted an improved reversible lithium storage capacity (1143 mA h g^{-1} at a current density of 100 mA g^{-1} after 200 cycles).

Molybdenum oxides are candidate anode materials for LIBs because MoO_3 exhibits good electrochemical properties, has a low cost, and is environmentally friendly [19–21]. Therefore, MoO_3 nanomaterials with diverse morphologies such as nanoparticles, hollow, nanobelts, naowiles, and porous structures have been prepared. Lee et al. [22] synthesized MoO_3 nanoparticles using hot filament chemical vapor deposition method (HFCVD) under an argon atmosphere. Zhao et al. [23] also synthesized MoO_3 hollow microspheres by a template-free solvothermal route and subsequent heat treatment in air. The MoO_3 hollow microspheres have a relatively high specific surface area. Chen et al. [24] prepared MoO_3 nanobelts by a hydrothermal method, in which the morphology of MoO_3 nanobelts was affected with the addition of PEG. MoO_{3-x} nanowires were prepared by Sunkara et al. [25] in a hot-filament chemical vapor deposition reactor. Ko et al. [26] prepared three-dimensional ordered macroporous structured MoO_3 by using a polystyrene bead template via ultrasonic spray pyrolysis.

In this study, low crystalline MoO_3/carbon composite microspheres, in which MoO_3 nanocrystals were distributed homogeneously in the amorphous C matrix, were directly prepared by one-step spray pyrolysis within several seconds. In here, MoO_3 was applied as the host material of carbon microspheres in this process because of its rich chemistry with multiple valence states, low electrical resistivity, high electrochemical activity toward lithium, and affordable cost. The resulting MoO_3/carbon composite microspheres worked better in terms of cycling and rate as anode materials for LIBs than did bare MoO_3 powders. The simple process introduced in this study is expected to be useful for the large-scale synthesis of TMOs/carbon composite microspheres as practical anode materials for LIBs. Furthermore, the synthesis strategy introduced is generally applied to synthesize various metal TMOs/carbon composites, including NiO, Co_3O_4, SnO_2, and Fe_2O_3, for a wide variety of applications including energy storage.

2. Materials and Methods

2.1. Sample Preparation

Low crystalline MoO_3/carbon composite microspheres, in which MoO_3 nanocrystals were distributed homogeneously in the amorphous C matrix, were directly prepared by a one-step spray pyrolysis. The spray pyrolysis system used in this study is shown in Figure S1. In brief, droplets were generated by a 1.7-MHz ultrasonic spray generator that consisted of six vibrators, and the droplets were carried to a quartz tube reactor (length = 1200 mm, diameter = 50 mm) by a flow of N_2 (flow rate = 5 L min^{-1}). The reactor temperature was fixed at 900 °C. The spray solution was prepared by dissolving 0.1 M of MoO_3 (98%, Sigma Aldrich, St. Louis, MO, USA), 12 g L^{-1} of polyvinylpyrrolidone (PVP, Mw 40,000, Daejung Chemicals and Metals, Siheung, Korea), and 0.02 M of sucrose in distilled water. Subsequently, an appropriate amount of hydrogen peroxide (30% H_2O_2, Sigma-Aldrich) was added to the above solution to obtain a clear spray solution. For the bare sample, MoO_3 powders without any carbon content were also prepared by spray pyrolysis. For this, the spray solution was prepared by dissolving 0.1 M of MoO_3 without a carbon precursor in H_2O_2 contained in distilled water. Subsequently, the spray pyrolysis was carried out with the prepared solution at a temperature of 900 °C by a flow of air (flow rate = 5 L min^{-1}).

2.2. Characterization Techniques

The microstructures of the resulting powders were observed by scanning electron microscopy (SEM; JEOL, JSM-6060, JEOL, Tokyo, Japan) and field-emission transmission electron microscopy (FE-TEM; JEOL, JEM-2100F, JEOL, Tokyo, Japan). The crystal phases were evaluated by X-ray diffractometry (XRD; X'Pert PRO MPD, PANalytical, Almelo, The Netherlands) using Cu K_{α} radiation (λ = 1.5418 Å). X-ray photoelectron spectroscopy (XPS; K-Alpha, Thermo Fisher Scientific, Waltham, MA, USA) with a focused monochromatic Al K_{α} at 12 kV and 20 mA was used to analyze the composition of the samples. A structural characterization of carbon in the sample was performed by Raman spectra (Jobin Yvon LabRam HR800, Horiba Jobin Yvon, Paris, France, excited by a 632.8 nm He–Ne laser) at room temperature. The surface areas of the powders were measured by the Brunauer–Emmett–Teller (BET) method, using N_2 as the adsorbate gas. Thermogravimetric analyses (TGA) were performed using a Pyris 1 TGA (Perkin Elmer, Waltham, MA, USA) within a temperature range of 25–650 °C and at a heating rate of 10 °C min^{-1} under a static air atmosphere.

2.3. Electrochemical Measurements

The electrochemical properties of the samples were analyzed by constructing a 2032-type coin cell. The lithium cell assembly was made in an Ar-filled glove box at room temperature where water and the oxygen concentration was kept at less than 1 ppm. The anode slurry was prepared by mixing the active material, carbon black, and sodium carboxymethyl cellulose (CMC) in a weight ratio of 7:2:1. The working electrodes were formed by coating the slurry onto copper foils and subsequently dried at 70 °C for 3 h. Li metal and a microporous polypropylene film were used as the counter electrode and the separator, respectively. The electrolyte was composed of 1 M $LiPF_6$ dissolved in a mixture of fluoroethylene carbonate/dimethyl carbonate (FEC/DMC; 1:1 v/v). The discharge/charge characteristics of the samples were investigated by cycling over a potential range of 0.001–3.0 V under CC (constant-current) conditions. Cyclic voltammograms were measured at a scan rate of 0.1 mV s^{-1}. The negative electrode measured 1.5 cm × 1.5 cm, and the mass loading of the active materials was kept at approximately 1.5 mg cm^{-2} in every electrochemical test. The electrochemical impedance spectra were obtained by performing alternating current electrochemical impedance spectroscopy (EIS; ZIVE SP1) over a frequency range of 0.01 Hz to 100 kHz.

3. Results and Discussion

Low crystalline MoO_3/C composite microspheres, in which MoO_3 nanocrystals were distributed homogeneously in the amorphous C matrix, were directly prepared by a one-step spray pyrolysis without any further treatment. Figure 1 shows the morphologies of the MoO_3/C composite microspheres obtained after the one-step spray pyrolysis. The powders were spherical and had diameters on the order of microns because they were formed from one droplet with several tens of micrometers by drying, decomposition, and crystallization inside the hot-wall reactor, as shown in Scheme 1. Additionally, there was no aggregation between the powders because the spray pyrolysis was carried out within a very short residence time of 6 s in a hot-wall reactor maintained at 900 °C under a N_2 atmosphere in Figure 1a,b. From a high-resolution TEM image in Figure 1c, it was hard to confirm the nanocrystal MoO_3 grains formed during spray pyrolysis in a microsphere structure because the amorphous-like, very small MoO_3 nanocrystals were formed during the spray pyrolysis at 900 °C within a short residence time of 6 s. The XRD result also showed the broad peak intensities of the β-MoO_3 phase in Figure 1d. The mean crystallite size of the MoO_3 powders, which was calculated from the width of the (011) peak using Scherrer's equation, was 4 nm. Grain growth of the MoO_3 nanocrystals was effectively prohibited both by the short residence time of the droplet in the reactor and by being surrounded by the carbon formed by the decomposition of PVP and sucrose during the process. The elemental mapping images shown in Figure 1e exhibited a homogeneous distribution of Mo, O, and C, which implies that the ultrafine MoO_3 nanocrystals were homogeneously composited with C in the microsphere structure.

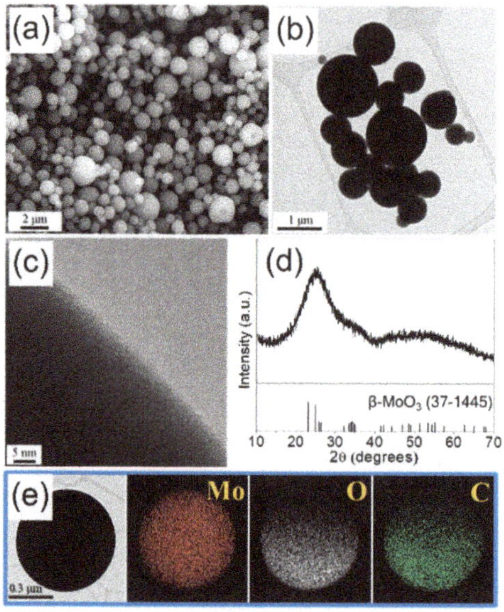

Figure 1. The (**a**) SEM, (**b**) TEM, (**c**) high-resolution TEM images, (**d**) XRD pattern, and (**e**) elemental mapping images of MoO_3/C composite microspheres.

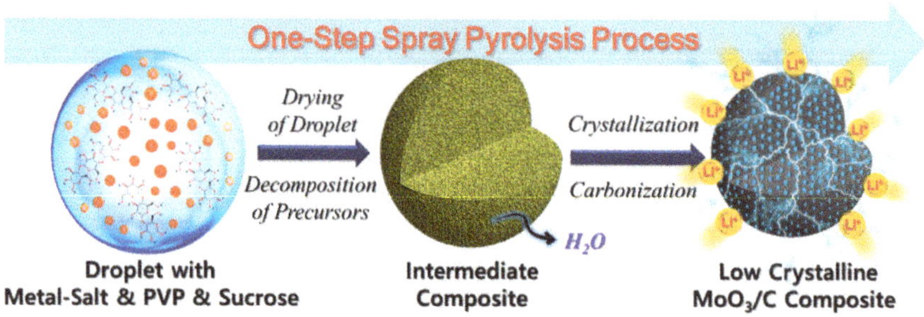

Scheme 1. The formation mechanism of the low crystalline MoO_3/C composite microspheres by the one-step spray pyrolysis process.

To identify the chemical composition of the MoO_3/C composite microspheres, XPS analysis was carried out, as shown in Figure 2. The XPS survey spectrum of the composite microspheres confirmed the presence of Mo, O, and C, as shown in Figure 2a. In the Mo 3d spectrum of the microspheres (Figure 2b), the main peaks occurred at binding energies of 231.7/232.7 eV for Mo 3d5/2 and 234.7/235.7 eV for Mo 3d3/2; the peaks located at 232.7 and 235.7 eV are characteristic of typical values of the 3d orbital doublet Mo^{6+}, and the minor ones centered on 231.7 and 234.7 eV corresponded to the 3d orbital doublet Mo^{5+}, which indicated that dangling bond sites where charges could be trapped existed in MoO_3 [27,28]. The C 1s XPS peak observed at 284.6 eV in Figure 2d corresponds to the binding energy of the sp^2 C–C bond of the carbon matrix [29–31].

The carbon matrix of the MoO_3/C composite microspheres was characterized by means of Raman spectroscopy. The degree of graphitization of the carbon material can typically be evaluated according to the intensity ratio of the D and G bands of carbon at approximately 1350 and 1590 cm^{-1}, respectively [32,33]. The peak intensity ratio between the D and G bands (I_D/I_G) for the MoO_3/C composite microspheres was approximately 3.2, and the absence of the 2D band at approx. 2685 cm^{-1} demonstrated that the carbon formed in the composite was fairly disordered. Thus, a large amount of the amorphous carbon was formed by the decomposition of both PVP and sucrose during the spray pyrolysis. In general, amorphous carbon has more capacity as an anode for LIBs than graphitic carbon, which is mainly contributed by pores and voids in the microcavities of the structure. The Thermogravimetric (TG) curve of the MoO_3/C composite microspheres in Figure 3b revealed a weight loss between 380 and 460 °C because of the degradation of amorphous carbon. Therefore, the content of amorphous carbon of the MoO_3/C composite microspheres estimated from the TG analysis was 26 wt %.

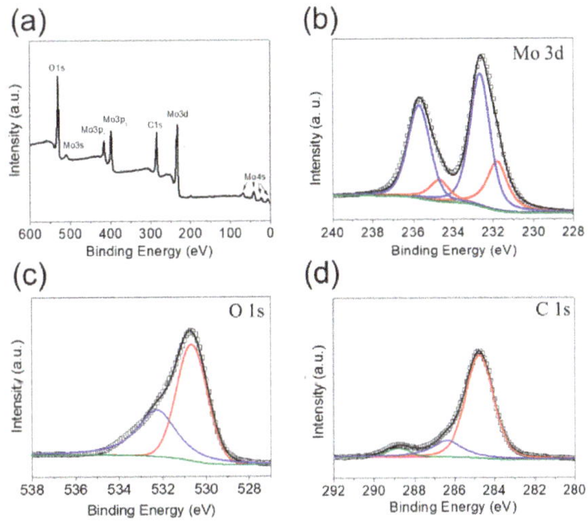

Figure 2. The XPS spectra of the MoO_3/C composite microspheres: (**a**) the survey XPS spectrum and high-resolution XPS spectra of (**b**) Mo 3d, (**c**) O 1S, and (**d**) C 1s.

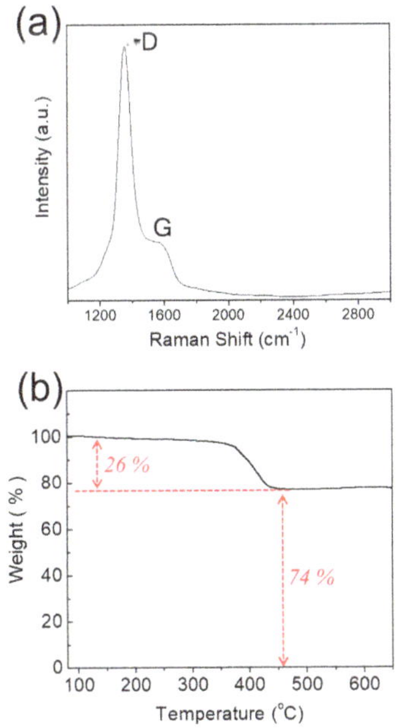

Figure 3. (**a**) The Raman spectrum and (**b**) thermogravimetric analysis (TGA) curve of the MoO_3/C composite microspheres.

In order to clearly prove the structural merits of MoO_3/C composite microspheres as anodes for Li^+ ion storage properties, bare MoO_3 powders without C were also prepared from the spray solution without either PVP and sucrose by spray pyrolysis, as shown in Figure 4. The mean particle size of the resulting bare MoO_3 powders, as measured from the SEM and TEM images in Figure 4a,b, was 420 nm and had no aggregation between the powders. Additionally, the resulting powders were angular, which is attributed to the crystal growth of MoO_3 particles because there was no carbon surrounding the particles during spray pyrolysis to prevent the growth of MoO_3 crystals during the short residence reaction time of the droplets. The high-resolution TEM image in Figure 4c shows clear lattice fringes separated by 0.23 nm, which corresponds to the (011) crystal plane of β-MoO_3 (JCPDS card No. 37–1445) [34]. The XRD pattern of the bare MoO_3 powders (Figure 4d) shows that they have different allotropes of MoO_3 structures, with no impurities. The thermodynamically favored α-MoO_3 phase was newly formed along with β-MoO_3 in the bare MoO_3 powders during spray pyrolysis. Bare MoO_3 powders without C were further confirmed by the elemental mapping images in Figure 4e. The BET surface areas of the MoO_3/C composite microspheres and of the bare MoO_3 powders were 4.3 and 0.6 $m^2\ g^{-1}$, respectively, in Figure S2.

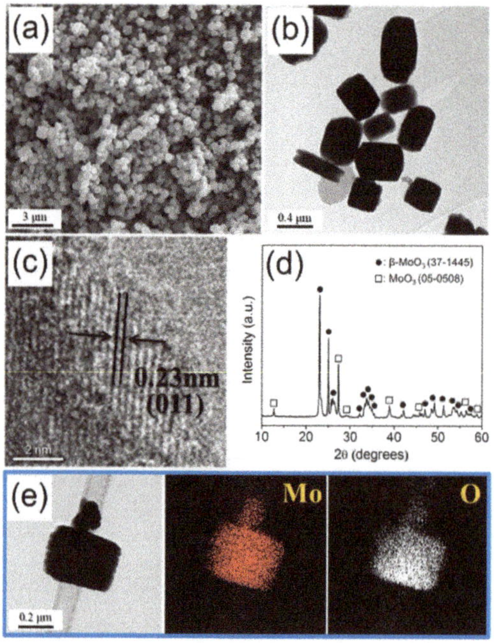

Figure 4. The (**a**) SEM, (**b**) TEM, (**c**) high-resolution TEM images, (**d**) XRD pattern, and (**e**) elemental-mapping images of the bare MoO_3 powders.

The electrochemical properties of the MoO_3/C composite microspheres are compared with those of the bare MoO_3 powders in Figure 5. The cyclic voltammogram (CV) curves of the MoO_3/C composite microspheres and bare MoO_3 powders performed in the 0.01–3.0 V range at a scanning rate of 0.01 mV s^{-1} for the first four cycles are shown in Figure 5a. In the first cathodic scan of the MoO_3/C composite microspheres, the broad peaks located at 1.16 V and 0.21 V are assigned to the interaction of Li^+ ions with the amorphous carbon matrix of the MoO_3/C composite and conversion reaction of Li_xMoO_3 to Mo_0 and Li_2O [35–37]. The peak at 0.05 V is also observed, caused by the Li^+ ion's intercalation into the C matrix [38,39]. In the anodic scans of the MoO_3/C composite microspheres, reversible peaks at 1.42 and 1.77 V are attributed to the monoclinic-orthorhombic-monoclinic phase transitions in the partially lithiated Li_xMoO_2 [35–37]. In the subsequent cycles, two redox peak pairs appeared at 0.21/1.3 and 1.42/1.77 V, which corresponded to the redox reaction of MoO_3 [35,40,41]. The bare MoO_3 powders showed peaks at 2.03 and 1.8 V in the first cathodic scan, which correspond to the generation of Li_xMoO_3, causing an irreversible structural change from the α-MoO_3 additionally formed in the bare MoO_3 powders to an amorphous phase [40–42]. The subsequent peak at 0.17 V results from the conversion reaction of Li_xMoO_3 to Mo_0 and Li_2O [35–37,41].

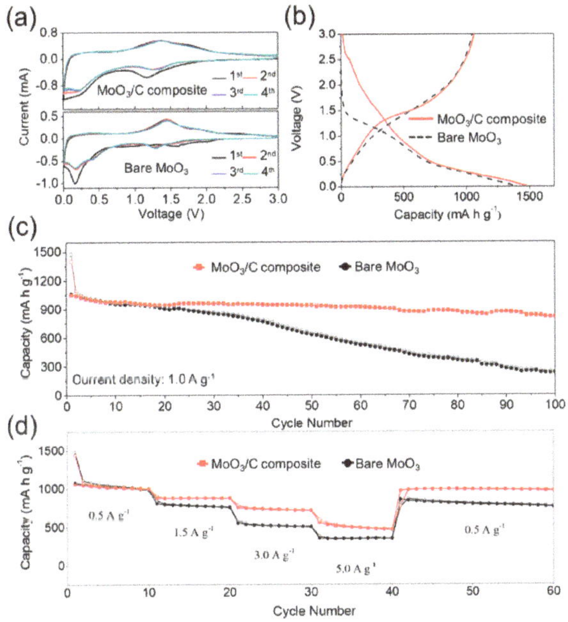

Figure 5. The electrochemical properties of the MoO_3/C composite microspheres and the bare MoO_3 powders: (**a**) CV curves, (**b**) charge-discharge curves, (**c**) cycling performances, and (**d**) rate performances.

The initial discharge-charge curves of the two samples at a current density of 1.0 A g^{-1} are shown in Figure 5b. The initial discharge capacities of the MoO_3/C composite microspheres and the bare MoO_3 powders were 1403 mA h g^{-1} and 1478 mA h g^{-1}, respectively, and their initial Coulombic efficiencies were 75% and 72%, respectively. Although the MoO_3/C composite microspheres contained C with a high irreversible capacity loss, the initial Coulombic efficiency of the MoO_3/C composite microspheres was relatively higher than that of the bare MoO_3 powders. The high structural damage to the bare MoO_3 powders in the first discharge and charge processes resulted in a low initial Coulombic efficiency. The discharge capacity and cycling properties of the MoO_3/C composite microspheres and bare MoO_3 powders at a current density of 1.0 A g^{-1} are shown in Figure 5c. Compared with bare MoO_3 powders, the MoO_3/C composite microspheres exhibited a satisfactorily stable cycling performance. The discharge capacity of the MoO_3/C composite microspheres decreased slightly from 1066 mA h g^{-1} (533 mA h cc^{-1}) to 808 mA h g^{-1} (404 mA h cc^{-1}) from the 2nd cycle to the 100th cycle, whereas that of the bare MoO_3 powders decreased rapidly from 1090 mA h g^{-1} (621 mA h cc^{-1}) to 239 mA h g^{-1} (136 mA h cc^{-1}) in the same cycle range. Additionally, the Coulombic efficiency of the MoO_3/C composite microspheres increased quickly to above 99% after the second cycle. The amorphous C matrix of MoO_3/C composite microspheres more effectively buffered the large volume change of the MoO_3 active material during the fast charging–discharging process. On the other hand, the structural destruction of the bare MoO_3 powders during repeated Li^+-ion insertion and desertion processes resulted in capacity fading continuously. Therefore, better cycling of the MoO_3/C composite microspheres could be achieved because of the improved structural stability of the MoO_3.

In order to evaluate the rate performances of both samples, electrochemical tests were performed at various current densities, as shown in Figure 5d. As the current densities increased from 0.5 to 1.5, 3.0, and 5.0 A g^{-1}, the MoO_3/C composite microspheres exhibited reversible discharge capacities of 999, 875, 716, and 467 mA h g^{-1}, respectively. However, the bare MoO_3 powders delivered a low reversible discharge capacity of 352 mA h g^{-1} at 5.0 A g^{-1} as shown in Figure 5d. The C matrix of the MoO_3/C composite microspheres improved the electrical conductivity of the sample. Additionally, the small, amorphous MoO_3 nanograins imbedded within the C matrix decreased the diffusion distance and increased the diffusion rate of the Li^+ ions, thus synergistically speeding up the rate of the MoO_3/C composite microspheres more than that of the bare MoO_3 powders.

The superior Li^+-ion storage properties of the MoO_3/C composite microspheres were supported by EIS analysis, as shown in Figure 6 [43–45]. Nyquist plots of the samples before and after cycles were obtained by deconvolution with a Randle-type equivalent-circuit model (Figure 6d). The MoO_3/C composite microspheres and bare MoO_3 powders had similar charge-transfer resistance (Rct) values before cycling, as shown in Figure 6a. However, the cell with the MoO_3/C composite microspheres obtained after 100 cycles showed a lower Rct value of 42 Ω compared to that of 134 Ω for the bare MoO_3 powders, as shown in Figure 6b,c. The structural destruction of the bare MoO_3 powders during the repeated Li^+-ion insertion and desertion processes increased the Rct values significantly. On the other hand, the MoO_3 nanograins embedded within the amorphous C were not pulverized during the repeated cycles. Moreover, the C matrix served as fast and continuous transport pathways for electrons upon cycling because of its high electrical conductivity. The high structural stabilities of the MoO_3/C composite microspheres with high lithium-ion storage capacities resulted in low Rct values during cycling. The MoO_3/C composite microspheres with a high structural stability during repeated lithium insertion and desertion reactions showed excellent cycling and rate performance, as shown in Figure 5.

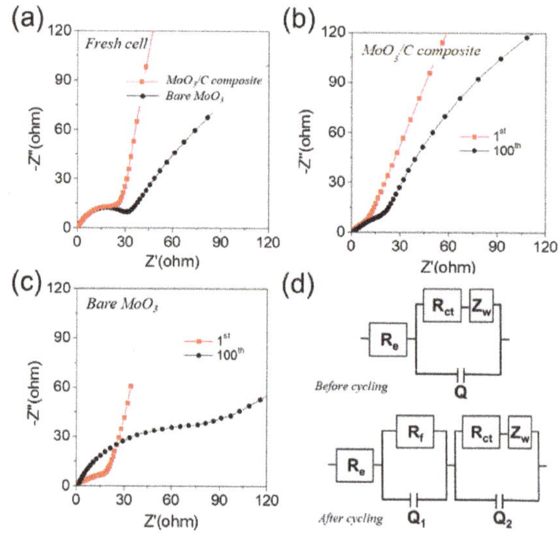

Figure 6. The impedance analysis of the MoO_3/C composite microspheres and the bare MoO_3 powders: (**a**) before cycling, (**b**) bare MoO_3 powders, (**c**) MoO_3/C composite microspheres, and (**d**) the tquivalent circuit model used for AC impedance fitting: R_{ct} = charge-transfer resistance, R_e = electrolyte resistance, R_f = SEI layer resistance, Q_1 = dielectric relaxation capacitance, and Q_2 = associated double layer capacitance.

The morphologies of the MoO_3/C composite microspheres and bare MoO_3 powders obtained after 100 cycles are shown in Figure 7. The bare MoO_3 powders were broken into several pieces after the cycles, as shown by the TEM image in Figure 7a. In contrast, the MoO_3/C composite microspheres maintained their morphologies quite well even after the repeated Li^+ insertion and desertion processes in Figure 7b,c. The excellent Li^+-ion storage properties of the MoO_3/C composite microspheres are, therefore, attributed to the improvement of the structural stability and electrical conductivity by the carbon composite.

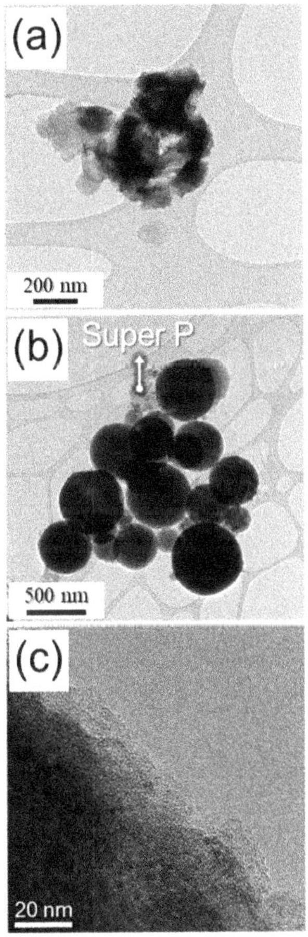

Figure 7. The morphologies of (**a**) bare MoO_3 powders and (**b**,**c**) MoO_3/C composite microspheres obtained after 100 cycles at a constant current density of 1.0 A g^{-1}.

4. Conclusions

In this study, low crystalline MoO_3/carbon composite microspheres in which MoO_3 nanocrystals were distributed homogeneously in the amorphous C matrix, were directly prepared by a one-step spray pyrolysis within several seconds. The MoO_3/carbon composite was spherical, with diameters on the order of microns, because they were formed from one droplet with several tens of micrometers by a series of drying, decomposition, and crystallization processes inside the hot-wall reactor during spray pyrolysis. The amorphous C matrix of the MoO_3/C composite microspheres effectively buffered the large volume change of the MoO_3 active material during the fast charging–discharging process.

Therefore, a better cycling of the MoO_3/C composite microspheres could be achieved because of the improved structural stability of the MoO_3. Additionally, the small MoO_3 nanograins imbedded within the C matrix decreased the diffusion distance and increased the diffusion rate of Li^+ ions, thus accelerating the rate of the MoO_3/C composite microspheres. The superior Li^+-ion storage properties of the MoO_3/C composite microspheres compared to those of the bare MoO_3 were supported by an EIS analysis and by observing the morphologies of the samples obtained after 100 cycles. The simple process introduced in this study is expected to be useful for the large-scale synthesis of TMOs/carbon composite microspheres for a wide variety of applications including energy storage.

Supplementary Materials: The following are available online at http://www.mdpi.com/2079-4991/9/4/539/s1, Figure S1: Schematic diagram of spray pyrolysis system applied in the preparation of MoO_3/C composite microspheres, Figure S2: N_2 adsorption-desorption isotherms measured at 77 K for the MoO_3/C composite microspheres and bare MoO_3 powders, Figure S3: Cycle properties of the MoO_3/C composite microspheres and the bare MoO_3 powders, Figure S4: TGA curve of the bare MoO_3 powders, Table S1: Fitted data obtained from the equivalent circuit for Nyquist plots.

Funding: This work was supported by the National Research Foundation of Korea (NRF) grant funded by the Korea government (MSIP) (NRF-2018R1A4A1024691, NRF-2017M1A2A2087577, and NRF-2018R1D1A3B07042514).

Conflicts of Interest: The authors declare no conflict of interest.

References

1. Chen, X.; Wang, Z.; Zhang, R.; Xu, L.; Sun, D. A novel polyoxometalate-based hybrid containing a 2D $[CoMo_8O_{26}]_\infty$ structure as the anode for lithium-ion batteries. *Chem. Commun.* **2017**, *53*, 10560–10563. [CrossRef]
2. Lin, H.B.; Rong, H.B.; Huang, W.Z.; Liao, Y.H.; Xing, L.D.; Xu, M.Q.; Li, X.P.; Li, W.S. Triple-shelled Mn_2O_3 hollow nanocubes: Force-induced synthesis and excellent performance as the anode in lithium-ion batteries. *J. Mater. Chem. A* **2014**, *2*, 14189–14194. [CrossRef]
3. Zhu, Z.; Wang, S.; Du, J.; Jin, Q.; Zhang, T.; Cheng, F.; Chen, J. Ultrasmall Sn nanoparticles embedded in nitrogen-doped porous carbon as high-performance anode for lithium-ion batteries. *Nano Lett.* **2013**, *14*, 153–157. [CrossRef]
4. Bai, Z.; Zhang, Y.; Zhang, Y.; Guo, C.; Tang, B.; Sun, D. MOFs-derived porous Mn_2O_3 as high-performance anode material for Li-ion battery. *J. Mater. Chem. A* **2015**, *3*, 5266–5269. [CrossRef]
5. Cui, L.; Shen, J.; Cheng, F.; Tao, Z.; Chen, J. SnO_2 nanoparticles@ polypyrrole nanowires composite as anode materials for rechargeable lithium-ion batteries. *J. Power Sources* **2011**, *196*, 2195–2201. [CrossRef]
6. Zhong, X.; Huan, H.; Liu, X.; Yu, Y. Facile synthesis of porous germanium-iron bimetal oxide nanowires as anode materials for lithium-ion batteries. *Nano Res.* **2018**, *11*, 3702–3709. [CrossRef]
7. Jin, L.; Qiu, Y.; Deng, H.; Li, W.; Li, H.; Yang, S. Hollow $CuFe_2O_4$ spheres encapsulated in carbon shells as an anode material for rechargeable lithium-ion batteries. *Electrochim. Acta* **2011**, *56*, 9127–9132. [CrossRef]
8. Wang, Y.; Xu, J.; Wu, H.; Xu, M.; Peng, Z.; Zheng, G. Hierarchical SnO_2–Fe_2O_3 heterostructures as lithium-ion battery anodes. *J. Mater. Chem.* **2012**, *22*, 21923–21927. [CrossRef]
9. Han, F.; Li, D.; Li, W.-C.; Lei, C.; Sun, Q.; Lu, A.-H. Nanoengineered polypyrrole-coated Fe_2O_3@C multifunctional composites with an improved cycle stability as lithium-ion anodes. *Adv. Funct. Mater.* **2013**, *23*, 1692–1700. [CrossRef]
10. Huang, X.H.; Tu, J.P.; Zhang, C.Q.; Chen, X.T.; Yuan, Y.F.; Wu, H.M. Spherical NiO-C composite for anode material of lithium ion batteries. *Electrochim. Acta* **2007**, *52*, 4177–4181. [CrossRef]
11. Jo, M.S.; Ghosh, S.; Jeong, S.M.; Kang, Y.C.; Cho, J.S. Coral-like yolk–shell-structured nickel oxide/carbon composite microspheres for high-performance Li-ion storage anodes. *Nano-Micro Lett.* **2019**, *11*, 3. [CrossRef]
12. Li, L.; Li, Z.; Fu, W.; Li, F.; Wang, J.; Wang, W. α-Fe_2O_3@C nanorings as anode materials for high performance lithium ion batteries. *J. Alloy. Compd.* **2015**, *647*, 105–109. [CrossRef]
13. Park, G.D.; Kim, J.H.; Park, S.-K.; Kang, Y.C. $MoSe_2$ embedded CNT-reduced graphene oxide composite microsphere with superior sodium ion storage and electrocatalytic hydrogen evolution performances. *ACS Appl. Mater. Interfaces* **2017**, *9*, 10673–10683. [CrossRef]

14. Cho, J.S.; Won, J.M.; Lee, J.-K.; Kang, Y.C. Design and synthesis of multiroom-structured metal compounds–carbon hybrid microspheres as anode materials for rechargeable batteries. *Nano Energy* **2016**, *26*, 466–478. [CrossRef]
15. Zhang, X.; Suresh Kumar, P.; Aravindan, V.; Liu, H.H.; Sundaramurthy, J.; Mhaisalkar, S.G.; Duong, H.M.; Ramakrishna, S.; Madhavi, S. Electrospun TiO_2–graphene composite nanofibers as a highly durable insertion anode for lithium ion batteries. *J. Phys. Chem. C* **2012**, *116*, 14780–14788. [CrossRef]
16. Bhaskar, A.; Deepa, M.; Narasinga Rao, T. MoO_2/multiwalled carbon nanotubes (MWCNT) hybrid for use as a Li-ion battery anode. *ACS Appl. Mater. Interfaces* **2013**, *5*, 2555–2566. [CrossRef] [PubMed]
17. Cho, J.S.; Hong, Y.J.; Kang, Y.C. Design and synthesis of bubble-nanorod-structured Fe_2O_3–carbon nanofibers as advanced anode material for Li-ion batteries. *ACS Nano* **2015**, *9*, 4026–4035. [CrossRef]
18. Park, S.-K.; Park, G.D.; Kang, Y.C. Three-dimensional porous microspheres comprising hollow Fe_2O_3 nanorods/CNT building blocks with superior electrochemical performance for lithium ion batteries. *Nanoscale* **2018**, *10*, 11150–11157. [CrossRef]
19. Whittingham, M.S. The role of ternary phased in cathode reactions. *Electrochem. Soc.* **1976**, *123*, 315–320. [CrossRef]
20. Choi, S.H.; Kang, Y.C. Crumpled graphene–molybdenum oxide composite powders: Preparation and application in lithium-ion batteries. *ChemSusChem* **2014**, *7*, 523–528. [CrossRef]
21. Xue, X.Y.; Chen, Z.H.; Xing, L.L.; Yuan, S.; Chen, Y.J. SnO2/a-MoO3 core-shell nanobelts and their extraordinarily high reversible capacity as lithium-ion battery anodes. *Chem. Commun.* **2011**, *47*, 5205–5207. [CrossRef] [PubMed]
22. Lee, S.H.; Kim, Y.H.; Deshpande, R.; Parilla, P.A.; Whitney, E.; Gillaspie, D.T.; Kim, M.J.; Mahan, A.H.; Zhang, S.B.; Dillon, A.C. Reversible lithium ion insertion in molybdenum oxide nanoparticles. *Adv. Mater.* **2008**, *20*, 3627–3632. [CrossRef]
23. Zhao, X.; Cao, M.; Hu, C. Thermal oxidation synthesis hollow MoO3 microspheres and their applications in lithium storage and gas-sensing. *Mater. Res. Bull.* **2013**, *48*, 2289–2295. [CrossRef]
24. Mohan, V.M.; Bin, H.; Chen, W. Enhancement of electrochemical properties of MoO3 nanobelts electrode using PEG as surfactant for lithium battery. *J. Solid State Electrochem.* **2010**, *14*, 1769–1775. [CrossRef]
25. Meduri, P.; Clark, E.; Kim, J.H.; Dayalan, E.; Sumanasekera, G.U.; Sunkara, M.K. MoO3–x nanowire arrays as stable and high-capacity anodes for lithium ion batteries. *Nano Lett.* **2012**, *12*, 1784–1788. [CrossRef] [PubMed]
26. Ko, Y.N.; Park, S.B.; Jung, K.Y.; Kang, Y.C. One-pot facile synthesis of ant-cave- structured metal oxide-carbon microballs by continuous process for use as anode materials in Li-ion batteries. *Nano Lett.* **2013**, *13*, 5462–5466. [CrossRef]
27. Xie, F.; Choy, W.C.H.; Wang, C.; Li, X.; Zhang, S.; Hou, J. Low-temperature solution-processed hydrogen molybdenum and vanadium bronzes for an efficient hole-transport layer in organic electronics. *Adv. Mater.* **2013**, *25*, 2051–2055. [CrossRef]
28. Zhu, Y.; Yuan, Z.; Cui, W.; Wu, Z.; Sun, Q.; Wang, S.; Kang, Z.; Sun, B. A cost-effective commercial soluble oxide cluster for highly efficient and stable organic solar cells. *J. Mater. Chem. A* **2014**, *2*, 1436–1442. [CrossRef]
29. Kim, J.H.; Oh, Y.J.; Kang, Y.C. Design and synthesis of macroporous $(Mn_{1/3}Co_{2/3})O$-carbon nanotubes composite microspheres as efficient catalysts for rechargeable Li-O_2 batteries. *Carbon* **2018**, *128*, 125–133. [CrossRef]
30. Park, G.D.; Lee, J.-K.; Kang, Y.C. Three-dimensional macroporous CNTs microspheres highly loaded with $NiCo_2O_4$ hollow nanospheres showing excellent lithium-ion storage performances. *Carbon* **2018**, *128*, 191–200. [CrossRef]
31. Liu, S.-Y.; Xie, J.; Pan, Q.; Wu, C.-Y.; Cao, G.-S.; Zhu, T.-J.; Zhao, X.-B. Graphene anchored with nanocrystal Fe_2O_3 with improved electrochemical Li-storage properties. *Int. J. Electrochem. Sci.* **2012**, *7*, 354–362.
32. Cho, J.S.; Lee, S.Y.; Kang, Y.C. First introduction of $NiSe_2$ to anode material for sodium-ion batteries: A hybrid of graphene-wrapped $NiSe_2$/C porous nanofiber. *Sci. Rep.* **2016**, *6*, 23338. [CrossRef] [PubMed]
33. Bhattacharya, K.; Gogoi, B.; Buragohain, A.K.; Deb, P. Fe_2O_3/C nanocomposites having distinctive antioxidant activity and hemolysis prevention efficiency. *Mater. Sci. Eng. C* **2014**, *42*, 595–600. [CrossRef]
34. Shakir, I.; Shahid, M.; Kang, D.J. MoO_3 and $Cu_{0.33}MoO_3$ nanorods for unprecedented UV/visible light photocatalysis. *Chem. Commun.* **2010**, *46*, 4324–4326. [CrossRef] [PubMed]

35. Sen, U.K.; Mitra, S. Synthesis of molybdenum oxides and their electrochemical properties against Li. *Energy Procedia* **2014**, *54*, 740–747. [CrossRef]
36. Yu, Z.; Jiang, H.; Gu, D.; Li, J.; Wang, L.; Shen, L. A new way to prepare MoO_3/C as anode of lithium ion battery for enhancing the electrochemical performance at room temperature. *J. Electrochem. Sci. Technol.* **2016**, *7*, 170–178. [CrossRef]
37. Zhou, J.; Lin, N.; Wang, L.; Zhang, K.; Zhu, Y.; Qian, Y. Synthesis of hexagonal MoO_3 nanorods and a study of their electrochemical performance as anode materials for lithium-ion batteries. *J. Mater. Chem. A* **2015**, *3*, 7463–7468. [CrossRef]
38. Ji, L.; Lin, Z.; Guo, B.; Medford, A.J.; Zhang, X. Assembly of carbon–SnO_2 core–sheath composite nanofibers for superior lithium storage. *Chem.-Eur. J.* **2010**, *16*, 11543–11548. [CrossRef]
39. Oh, S.H.; Kim, J.K.; Kang, Y.C.; Cho, J.S. Three-dimensionally ordered mesoporous multicomponent (Ni, Mo) metal oxide/N-doped carbon composite with superior Li-ion storage performance. *Nanoscale* **2018**, *10*, 18734–18741. [CrossRef] [PubMed]
40. Wu, D.; Shen, R.; Yang, R.; Ji, W.; Jiang, M.; Ding, W.; Peng, L. Mixed molybdenum oxides with superior performances as an advanced anode material for lithium-ion batteries. *Sci. Rep.* **2017**, *7*, 44697. [CrossRef] [PubMed]
41. Zhao, G.; Zhang, N.; Sun, K. Electrochemical preparation of porous MoO_3 film with a high rate performance as anode for lithium ion batteries. *J. Mater. Chem. A* **2013**, *1*, 221–224. [CrossRef]
42. Xia, Q.; Zhao, H.; Du, Z.; Zeng, Z.; Gao, C.; Zhang, Z.; Du, X.; Kulka, A.; Świerczek, K. Facile synthesis of MoO_3/carbon nanobelts as high-performance anode material for lithium ion batteries. *Electrochim. Acta* **2015**, *180*, 947–956. [CrossRef]
43. Lee, J.H.; Oh, S.H.; Jeong, S.Y.; Kang, Y.C.; Cho, J.S. Rattle-type porous Sn/C composite fibers with uniformly distributed nanovoids containing metallic Sn nanoparticles for high-performance anode materials in lithium-ion batteries. *Nanoscale* **2018**, *10*, 21483–21491. [CrossRef] [PubMed]
44. Ma, X.-H.; Feng, X.-Y.; Song, C.; Zou, B.-K.; Ding, C.-X.; Yu, Y.; Chen, C.-H. Facile synthesis of flower-like and yarn-like α-Fe_2O_3 spherical clusters as anode materials for lithium-ion batteries. *Electrochim. Acta* **2013**, *93*, 131–136. [CrossRef]
45. Wang, Y.; Xu, M.; Peng, Z.; Zheng, G. Direct growth of mesoporous Sn-doped TiO_2 thin films on conducting substrates for lithium-ion battery anodes. *J. Mater. Chem. A* **2013**, *1*, 13222–13226. [CrossRef]

© 2019 by the author. Licensee MDPI, Basel, Switzerland. This article is an open access article distributed under the terms and conditions of the Creative Commons Attribution (CC BY) license (http://creativecommons.org/licenses/by/4.0/).

nanomaterials

Article

Compact Integration of TiO_2 Nanoparticles into the Cross-Points of 3D Vertically Stacked Ag Nanowires for Plasmon-Enhanced Photocatalysis

Vo Thi Nhat Linh [1,2], Xiaofei Xiao [3], Ho Sang Jung [1], Vincenzo Giannini [3,4], Stefan A. Maier [3,5], Dong-Ho Kim [1,*], Yong-Ill Lee [2,*] and Sung-Gyu Park [1,*]

1 Advanced Nano-Surface Department (ANSD), Korea Institute of Materials Science (KIMS), Changwon, Gyeongnam 51508, Korea; vtnl94@kims.re.kr (V.T.N.L.); jhs0626@kims.re.kr (H.S.J.)
2 Department of Chemistry, Changwon National University, Changwon, Gyeongnam 51140, Korea
3 The Blackett Laboratory, Department of Physics, Imperial College London, London SW7 2AZ, UK; xiaofei.xiao15@imperial.ac.uk (X.X.); v.giannini@imperial.ac.uk (V.G.); s.maier@imperial.ac.uk (S.A.M.)
4 Instituto de Estructura de la Materia (IEM-CSIC), Consejo Superior de Investigaciones Científicas, Serrano 121, 28006 Madrid, Spain
5 Chair in Hybrid Nanosystems, Nanoinstitute Munich, Faculty of Physics, Ludwig-Maximilians-Universität München, 80539 München, Germany
* Correspondence: dhkim2@kims.re.kr (D.-H.K.); yilee@changwon.ac.kr (Y.-I.L.); sgpark@kims.re.kr (S.-G.P.); Tel.: +82-55-280-3632 (S.-G.P.)

Received: 27 February 2019; Accepted: 13 March 2019; Published: 20 March 2019

Abstract: The compact integration of semiconductor TiO_2 nanoparticles (NPs) into the 3D crossed region of stacked plasmonic Ag nanowires (NWs) enhanced the photocatalytic activities through synergistic effects between the strong localized surface plasmon resonance (LSPR) excitation at the 3D cross-points of the Ag NWs and the efficient hot electron transfer at the interface between the Ag NWs and the TiO_2 NPs. This paper explored new hybrid nanostructures based on the selective assembly of TiO_2 NPs onto 3D cross-points of vertically stacked Ag NWs. The assembled TiO_2 NPs directly contacted the 3D Ag NWs; therefore, charge separation occurred efficiently at the interface between the Ag NWs and the TiO_2 NPs. The composite nanomaterials exhibited high extinction across the ultraviolet-visible range, rendering the nanomaterials high-performance photocatalysts across the full (ultraviolet-visible) and the visible spectral regions. Theoretical simulations clearly revealed that the local plasmonic field was highly enhanced at the 3D crossed regions of the vertically stacked Ag NWs. A Raman spectroscopic analysis of probe dye molecules under photodegradation conditions clearly revealed that the nanogap in the 3D crossed region was crucial for facilitating plasmon-enhanced photocatalysis and plasmon-enhanced spectroscopy.

Keywords: 3D hybrid nanostructures; localized surface plasmon resonance; hot electrons; environmental remedy; plasmon-enhanced photocatalysis

1. Introduction

Semiconductor photocatalysts have been widely investigated for use in environmental science and technology applications, especially toward the degradation of organic pollutants [1–4], the photocatalytic production of hydrogen [5–7], and the photocatalytic reduction of carbon dioxide [8–10]. However, conventional semiconductor photocatalysts exhibit poor photocatalytic performances under visible light illumination ($\lambda > 400$ nm) due to their wide band gaps. For example, titanium dioxide (TiO_2) in the crystalline anatase phase has a band gap of 3.2 eV ($\lambda = 388$ nm), and TiO_2 semiconductor nanomaterials are active under ultraviolet (UV) light (over only 7.5% of the full solar spectrum). Around 54% of solar power falls within the visible and near-infrared regions (from 400 nm to

800 nm); therefore, significant efforts have been applied toward increasing the photoresponses of semiconductors in this spectral region and enhancing photocatalytic performances.

Recently, hybrid nanostructures containing semiconductor and plasmonic metal nanomaterials have been used to enhance photocatalytic activities over the visible range. For example, integrating TiO_2 nanoparticles (NPs) with plasmonic nanostructures can significantly improve photocatalysis performances by enhancing the localized surface plasmon resonance (LSPR) of plasmonic nanostructures, enabling efficient hot electron transfer to TiO_2 NPs [11–16]. At the LSPR excitation, plasmonic nanostructures can absorb and concentrate visible light at nanoscale gaps (hot spots) between metallic nanostructures, and highly energetic hot electrons can be injected into nearby TiO_2 NPs. As a result of this plasmonic sensitization process, a wide band gap TiO_2 material that is inactive under visible light can become active under visible light [11–17].

Here, we report the development of 3D hybrid nanostructures through the compact integration of TiO_2 NPs into the 3D cross-points of vertically stacked Ag NWs. The 3D hybrid nanostructures were prepared using a simple two-step vacuum filtration process, applied first to Ag NWs and second to TiO_2 NPs, over microfiber filters. We observed that low concentrations of TiO_2 NPs selectively localized at the small nanogap region of the 3D crossed Ag NWs. The resulting 3D composite nanomaterials exhibited a noticeable absorption across the entire UV and visible region due to a high density of hot spots in the 3D stacked Ag NWs [18]. At optimized concentrations of the TiO_2 NPs and Ag NWs, the photocatalytic efficiencies were measured to be 49.8% over 10 min illumination and 91.3% over 60 min illumination under standard air mass (AM) 1.5G conditions. We performed theoretical simulations of the local field enhancements in the 3D crossed regions using different wavelengths, and we systematically investigated the Raman spectral changes of organic dyes as a function of the photodegradation conditions. The results clearly revealed the crucial contribution of the hot spots (3D cross-points) to the highly enhanced photocatalytic activities and surface-enhanced Raman spectroscopy (SERS).

2. Materials and Methods

2.1. Fabrication of the 3D Hybrid Nanostructures

The hybrid nanostructures were fabricated using a two-step vacuum filtration process over a glass microfiber filter (HG00047F, HyundaiMicro, Tokyo, Japan). First, 4 mL of a 0.5 wt% Ag NWs aqueous solution (N&B Co, Ltd., Daejeon, Korea) was poured onto the microfiber filter, then 4 mL of a colloidal solution containing TiO_2 NPs was applied onto the 3D stacked Ag NWs without breaking the vacuum. The composite substrate was dried on a hot plate at 150 °C for 3 min. The TiO_2 NPs aqueous solution (Sigma Aldrich, St. Louis, Missouri, MO, USA) used in our experiments was a mixture of the anatase (80%) and rutile (20%) phases.

2.2. Photodegradation of Methylene Blue

Each substrate was immersed into a petri dish containing 10 mL of a 0.05 mM MB aqueous solution and remained on the bottom of the petri dish under illumination generated from a Xenon solar simulator lamp (AM 1.5G, 100 mW/cm^2) as a UV-visible light source. A UV-cutoff filter was placed between the lamp and the petri dish during photocatalytic activity testing in the visible range ($\lambda > 400$ nm). The photocatalytic activities of the TiO_2 NPs themselves were tested by mixing a TiO_2 colloidal solution with an MB aqueous solution and measuring photodegradation using a solar simulator.

2.3. Characterizations

The surface morphologies of the 3D nanostructures were characterized by field emission scanning electron microscopy (FE-SEM; JSM-6700F, Joel, Tyoko, Japan) and transmission electron microscopy (TEM; JEM-2100F, Joel, Tyoko, Japan), respectively. Diffuse reflectance spectra were measured using a UV-Vis-NIR spectrophotometer (Cary 5000, Agilent Technology, Santa Clara, CA, USA). Visible light is mainly reflected by glass microfibers; therefore, the extinction spectra were converted directly from the diffuse reflectance spectra. The SERS spectra were measured using a handheld Raman spectrometer (CBEx, Snowy Range Instruments, Laramie, WY, USA) equipped with 785 nm and 638 nm lasers. The incident laser power and exposure time were 10 mW and 0.01 s for the SERS measurements, respectively.

2.4. Numerical Simulations

Numerical simulations were carried out using finite difference time domain analysis (FDTD) techniques. In these simulations, transverse magnetic polarized incident waves with different wavelengths (450 nm, 532 nm, 633 nm, and 785 nm, respectively) were applied to the crossed Ag NW nanostructure. The simulation was simplified by including three crossed Ag NWs (the upper and lower NWs were perpendicular to the central NW). The Ag NWs were modeled as five-sided equilateral pentagons, and the radius of the pentagon was fixed at 25 nm. The mesh size was set to 0.2 nm around the junction of the crossed Ag NWs. Perfectly matched layer (PML) formulation was applied in all directions. The complex permittivity of Ag was adopted from Reference [19].

3. Results and Discussion

3.1. Compact Integration of the TiO_2 NPs into the 3D Cross-Points of Vertically Stacked Ag NWs

3D hybrid TiO_2/Ag NW structures were fabricated using a simple two-step vacuum filtration method. First, 3D-stacked Ag NWs were prepared by vacuum filtering 4 mL of a dispersion containing 0.5 wt% Ag NWs over a glass microfiber filter [20,21]. Then, 4 mL of an aqueous solution containing TiO_2 NPs were poured over the 3D stacked Ag NWs without breaking the vacuum. Figure 1a shows a scanning electron microscopy (SEM) image of the 3D stacked Ag NWs. Within the 3D multilayered Ag NWs, small nanogaps formed at the 3D cross-points at which the NWs were vertically stacked. Transmission electron microscopy (TEM) images clearly revealed that each Ag NW assumed a pentagonal shape, and small nanogap regions formed at the 3D cross-points, at which the lower and upper layers were closely spaced (dotted rectangle in Figure 1b). As small pores formed in the 3D crossed region, 22 nm (average diameter) TiO_2 NPs were selectively filtered during the vacuum filtration process. Figure 1c,d show the 3D hybrid nanostructures consisting of TiO_2 NPs and 3D Ag NWs upon application of a 0.002 wt% TiO_2 NP solution to the 3D stacked Ag NWs. The small TiO_2 NPs were mainly filtered from the solution and remained on top of the 3D cross-points regions, resulting in the compact integration of TiO_2 NPs into the 3D hot spot region. These interfaces enhanced the interactions between Ag NWs and the TiO_2 NPs. The 3D stacked AgNWs could be also considered as photocatalytic membranes [22,23]. In addition, the compact hybrid nanostructures embedded on glass microfiber filter would be easily separated from treated solution, which can be applied in water treatment [24,25]. The TiO_2 NPs covered the entire 3D surface of the Ag NWs when a 0.01 wt% TiO_2 NP aqueous solution was used (Figure S1).

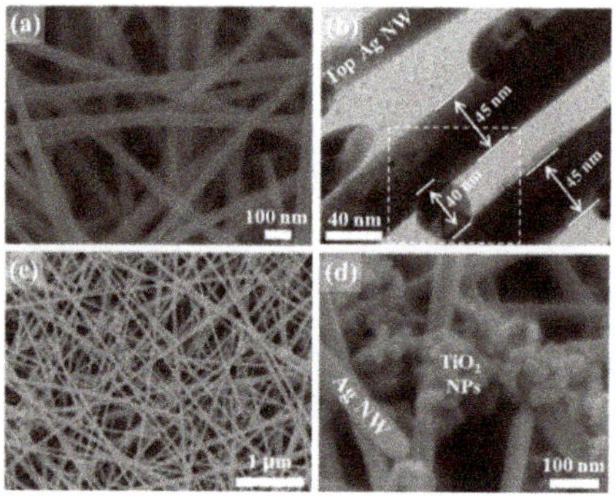

Figure 1. (**a**) SEM and (**b**) TEM images of 3D stacked Ag NWs. (**c**,**d**) SEM images of 3D hybrid nanostructures consisting of TiO_2 NPs and 3D Ag NWs. The image clearly shows the compact integration of TiO_2 NPs onto the crossed region of the 3D Ag NWs.

3.2. Optical Properties and Local Field Enhancements of the 3D Hybrid Nanostructures

Figure 2 presents the extinction spectra of 3D plasmonic Ag NWs and 3D TiO_2/Ag hybrid nanostructures prepared with different TiO_2 concentrations (see Figure S2 for the extinction spectra of the bare microfiber filters and TiO_2 NPs). The microfiber substrate deposited using 0.01 wt% TiO_2 NPs exhibited a very low extinction in the visible region but a very high UV extinction below 400 nm. These results indicated the presence of UV light–TiO_2 semiconductor interactions near the band gap of crystalline TiO_2 [26]. The green tangent line indicates 388 nm, the band gap energy (3.2 eV) of the anatase TiO_2 (Figure S2). The 3D Ag NWs exhibited strong interactions from the UV to the visible region (black line in Figure 2). The highest extinction near 326 nm was attributed to an interband transition (3.8 eV) of Ag. The 3D stacked Ag NWs also had a high extinction across the entire visible wavelength range. The 3D hybrid nanostructures possessed higher extinction than the 3D stacked Ag NWs in visible region. The high extinction of the 3D hybrid nanostructures agreed well with the high intensity of visible light under standard AM 1.5G solar illumination (navy line in Figure 2). Therefore, the 3D hybrid nanostructures were expected to increase the photocatalytic activities through enhanced light–matter interactions.

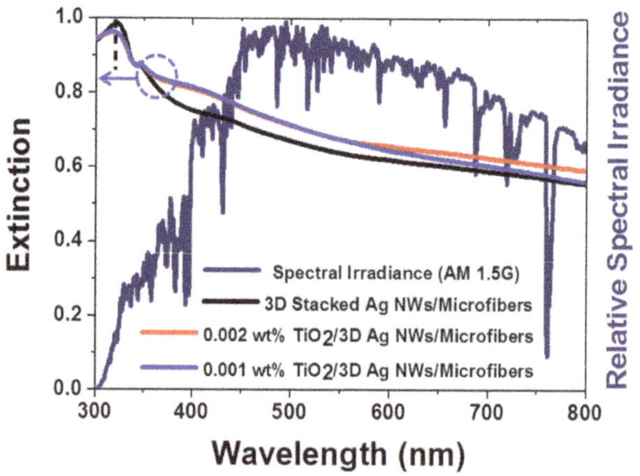

Figure 2. Extinction spectra of different nanomaterials and spectral irradiance of AM 1.5G. The composite nanostructures displayed high extinction in the visible range, in agreement with the solar spectrum in the visible range. The black lines indicate the interband transitions of Ag.

Another prominent property of the 3D plasmonic or hybrid nanostructures is the local field enhancement due to a coupled LSPR effect at the 3D cross-points [27–31]. Figure 3 shows the near-field enhancement distribution contours of orthogonal Ag NWs (similar to the Ag NW stack indicated by the white dotted area in Figure 1b) for different incident wavelengths obtained from finite-difference time-domain (FDTD) simulations (Figure S3). The strong field was localized and confined at the nanogap region of the cross-points of vertically stacked Ag NWs, regardless of the wavelength of incident light. The field intensity enhancement could be further increased by increasing the curvature of the metal nanostructures due to a lightning rod effect, indicating that the pentagonal shape provided stronger enhancement than the circular shape [32]. The maximum field enhancement was obtained at the bottom corners of the central pentagonal Ag NW. The maximum field enhancement increased by an order of magnitude at shorter wavelengths (Table 1). The average field enhancement along the middle Ag NW surface could be extracted from the simulation results (Table 1). Note that we extracted the field enhancement value 0.5 nm away from the Ag NW surface to avoid any staircase effects in the simulation (green lines in Figure S4). The average field enhancement corresponded to the maximum field enhancement (Table 1). The numerical simulations suggested that a highly enhanced local field could be generated at the cross-points of the 3D stacked Ag NWs. Therefore, it is important to deposit TiO_2 NPs selectively onto these hot spot regions to ensure efficient charge separation and enhanced photocatalysis.

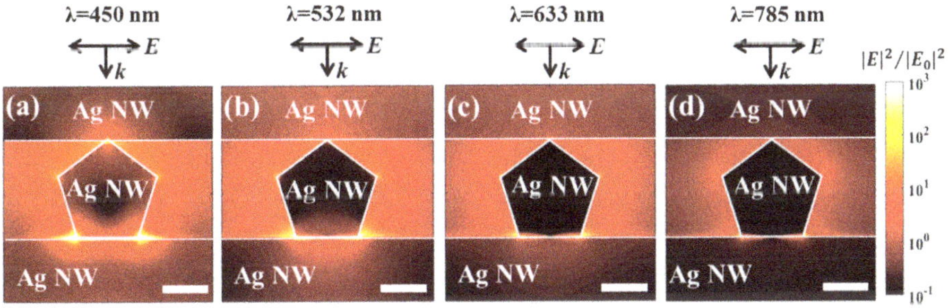

Figure 3. Finite difference time domain (FDTD) simulations of the electric field distribution ($|E|^2/|E_0|^2$) at the junctions of crossed Ag NWs, using the wavelengths (**a**) 450 nm, (**b**) 532 nm, (**c**) 633 nm, and (**d**) 785 nm. The k and E vectors in the figure show the incident direction and the polarization direction of the illumination, respectively. The white lines are used to indicate the boundaries of Ag NWs. All scale bars indicate 20 nm.

Table 1. Maximum and average field enhancement of crossed Ag NWs at different wavelengths.

Wavelength (nm)	Maximum Field Enhancement ($|E|^2/|E_0|^2$)	Average Field Enhancement ($|E|^2/|E_0|^2$)
450	2.9×10^4	5.6×10^2
532	3.2×10^3	1.0×10^2
633	4.3×10^2	1.8×10^1
785	8.6×10^1	4.4×10^0

3.3. *Photocatalytic Performance*

The photocatalytic activities of various nanomaterials were examined using the photodegradation of an organic dye (methylene blue (MB)). The photocatalytic degradation was carried out under AM 1.5G (or 1 sun) simulated illumination with an intensity of 100 mW/cm^2, which are standard light illumination conditions for measuring solar cell efficiency [33]. This standard light illumination provided a quantitative photocatalyst performance comparison, unlike other light sources, such as high-pressure mercury lamps with a high power [34]. All photocatalysts were independently examined in the presence of 10 mL of a 0.05 mM MB aqueous solution under 10 min illumination. Figure 4a shows the absorbance changes of an MB solution after 10 min light illumination of each photocatalytic substrate. The main absorbance peak of MB, at 665 nm, was found to decrease in intensity after light illumination of each photocatalyst. Negligible MB degradation was observed during 10 min illumination in the absence of a substrate, indicating that MB itself did not decompose under illumination. Interestingly, the 3D stacked plasmonic Ag NWs exhibited higher photocatalytic activities compared to the TiO_2 NPs, primarily due to LSPR-induced hot electron generation and the direct decomposition of MB [35,36]. The 3D hybrid nanostructures composed of 0.002 wt% TiO_2 NPs and 3D stacked Ag NWs showed the highest solution MB absorbance changes in the UV-visible and visible ranges (Figure 4b). At higher concentrations (i.e., 0.01wt% TiO_2 NP solution) of TiO_2, the photocatalytic activities decreased due to inefficient charge separation at the Ag NWs and TiO_2 interface and blockage of MB molecular mass transport into the hot spot regions by the large surface coverage of TiO_2 NPs (Figure S1). The photocatalytic activities of the different nanostructures were examined by calculating the decrease in MB absorbance after illumination (Table 2). The TiO_2 photocatalyst was active only under UV illumination due to its large band gap. No MB degradation was observed under visible illumination in the presence of the TiO_2 photocatalyst (Figure S5). The stacked 3D Ag NWs showed better activities in both the UV-visible and visible ranges compared to the TiO_2 NPs. The optically excited plasmonic nanostructures activated the chemical transformations directly on their surfaces [37–40]. Among the TiO_2/Ag NW composite nanomaterials, the composite prepared with 0.002 wt% TiO_2 NPs exhibited the best MB photocatalytic activities under both spectral ranges. In the presence of the 3D hybrid nanostructures, the MB concentration degraded to one-half of its original value after only 10 min illumination.

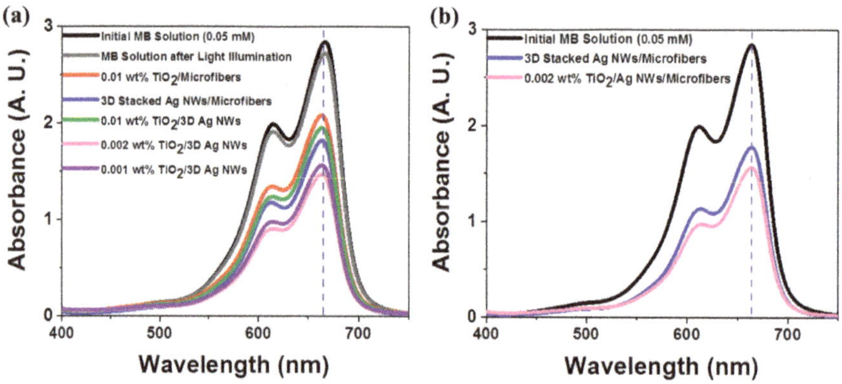

Figure 4. Photocatalytic performances of 0.05 mM methylene blue (MB) aqueous solutions after 10 min light illumination in the presences of different photocatalysts under (**a**) AM 1.5G simulated illumination and (**b**) visible and near-infrared illumination (λ > 400 nm).

The 3D composite nanostructures composed of 0.002 wt% TiO_2 and 3D Ag NWs were examined in time-dependent MB degradation studies under different light illumination conditions. The changes in the UV-visible spectra during photodegradation are shown in Figure 5. The main absorption peak of MB at 665 nm decreased markedly with the irradiation time (up to 60 min) and almost vanished after 90 min (Figure 5a). After 90 min of irradiation, the blue color of the initial MB solution became transparent due to the photodecomposition of MB molecules. We observed almost the same photocatalytic performance in the presence of the 3D composite under visible light illumination (λ > 400 nm), as shown in Figure 5b. The 3D composite performed effectively under both full spectral and visible spectral (λ > 400 nm) illumination.

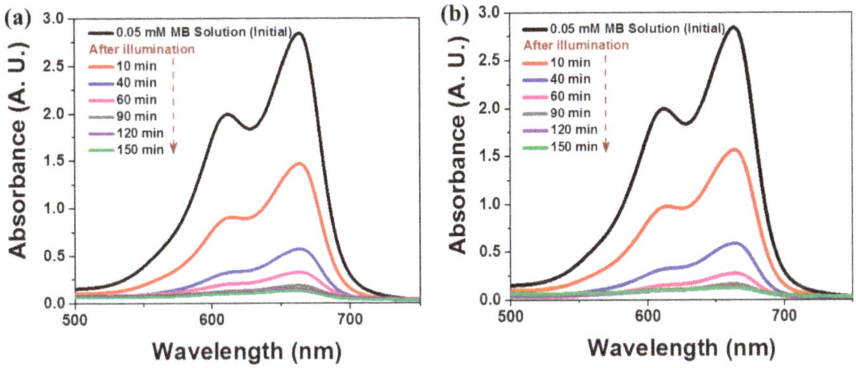

Figure 5. Time-dependent UV-vis spectra of an MB aqueous solution after illumination in the presence of 0.002 wt% TiO_2/Ag NW composites under (**a**) UV-visible light and (**b**) visible light.

Table 2. The photocatalytic efficiencies of different photocatalysts under AM 1.5G and visible and NIR (the UV was cut off by a filter) light after 10 min illumination*.

Photocatalysts	UV-Visible	Visible
TiO_2 0.01 wt%/Microfibers*	14.3%	0.2%
3D Stacked Ag NWs/Microfibers*	36.3%	37.8%
0.01 wt% TiO_2/3D Ag NWs Composite*	32.4%	23.7%
0.002 wt% TiO_2/3D Ag NWs Composite*	49.8%	46%
0.001 wt% TiO_2/3D Ag NWs Composite*	46.7%	45.8%

* 10 mL 0.05 mM MB was used to test the photodegradation of organic dyes during illumination.

The photocatalytic efficiencies of the 3D composite nanomaterials are illustrated in Figure 6. C_0 is the initial concentration of MB, and C is its concentration after each irradiation time. About 90% of the initial MB concentration was photodegraded under UV/visible and visible illumination after 60 min. The rate of MB degradation decreased after 60 min. This decrease was attributed to the fact that the substrate was placed at the bottom of a petri dish, and the MB molecules in the aqueous solution were randomly dispersed throughout the 10 mL volume during illumination. The dilution of intact MB molecules decreased the rate of collision between the organic molecules and the active site. The diffusion-limited behavior of the dye molecules required longer illumination times to achieve total decomposition of the organic dyes. Due to the diffusion-limited behavior, we divided the photocatalytic kinetics into two stages: before 60 min and after 60 min. The photodegradation process was assumed to follow Langmuir–Hinshelwood kinetics, $ln\ (C/C_0) = -kt$, where C is the MB concentration as a function of the photodegradation time (t), C_0 is the initial concentration, and k is the reaction rate constant (insets of Figure 6) [41,42]. The rate constant obtained under UV-visible illumination was measured to be 3.893×10^{-2} min^{-1} during the initial stage and decreased to 1.438×10^{-2} min^{-1} after 60 min (Inset of Figure 6a). Under visible light illumination, the values of k were found to be 3.876×10^{-2} min^{-1} and 1.084×10^{-2} min^{-1}, respectively (inset of Figure 6b).

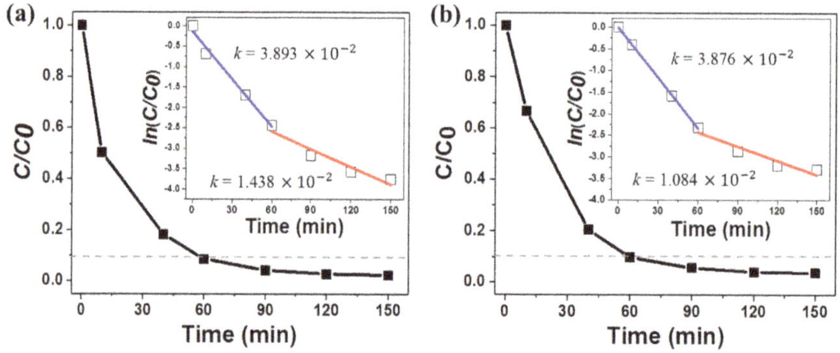

Figure 6. Photodegradation of an MB aqueous solution in the presence of 0.002 wt% TiO_2/Ag NW composite nanomaterials under (**a**) UV-visible light and (**b**) visible light. The inset figures show the linear fits to $ln(C/C_0)$ vs. time.

3.4. Mechanism Underlying the Plasmon-Enhanced Photocatalysis

Semiconductor TiO_2 NPs are typically integrated with 3D plasmonic Ag NWs with the goal of using the LSPR properties of 3D plasmonic Ag NWs to enhance visible light absorption and channel hot electrons to the TiO_2 NPs for efficient photocatalysis, a plasmonic sensitization process [17]. In 3D hybrid nanostructures, a Schottky barrier forms at the junction interface between Ag and TiO_2 (Figure 7). This barrier blocks electron transfer from the Ag NWs to the TiO_2 NPs. Upon excitation of the LSPR, highly energetic electrons (hot electrons) have sufficient energy to overcome the Schottky barrier, and they can inject into the conduction band of the TiO_2. The plasmonic sensitization process shuttles extra electrons to the wide-band gap (3.2 eV) TiO_2, which is originally inactive under visible light. The TiO_2 can then perform catalytic reduction reactions [5,17]. The holes left behind in the Ag NWs can be used to oxidize organic molecules. As most of the electron/hole pairs are produced in the field-enhanced region of the 3D crossing points of vertically stacked Ag NWs (Figure 3), efficient charge separation occurs at the interface between the 3D cross-points and nearby TiO_2 NPs. Therefore, the compact integration of crystalline TiO_2 NPs into the 3D cross-points can enhance the photocatalytic performance over the full spectral range.

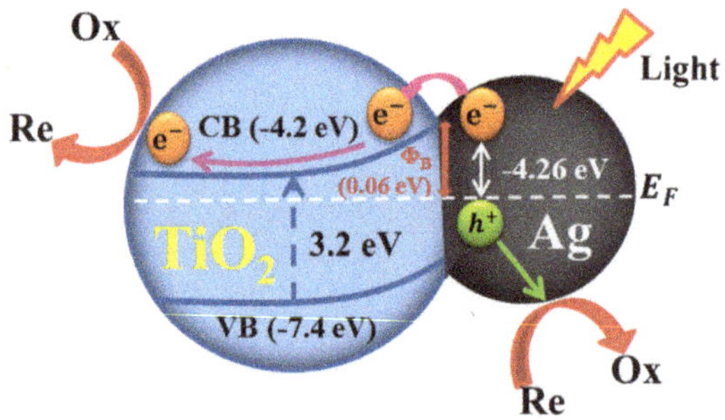

Figure 7. Schematic band energy diagram of charge transfer at the Ag NW/TiO_2 NP interface (not drawn to scale). CB and VB indicate the conduction band and valence band of anatase TiO_2. E_F presents the Fermi level of Ag. Φ_B is the Schottky barrier height.

3.5. Spectroscopic Investigation of Plasmon-Enhanced Photocatalysis

We collected the Raman spectra of organic dye molecules to investigate their plasmon-enhanced photocatalytic degradation by 3D stacked Ag NWs and 3D hybrid nanostructures. First, 3 μL of a 5 μM MB aqueous solution was applied to each photocatalytic substrate. After drying, SERS spectra were obtained using a portable Raman spectrometer [21,43]. The SERS measurements were collected using a low-power 10 mW illumination beam with a spot size of 20 μm incident on each substrate for a very short laser exposure time of 0.01 s to avoid photocatalytic degradation of the MB molecules. The laser was then directed onto the same spot on the substrate for 10 s to induce photocatalytic degradation of the MB molecules. SERS measurements were performed to detect any SERS intensity changes in the probe molecules. Each cycle consisted of a photodegradation process (10 s illumination) and a SERS measurement (0.01 s illumination of 10 mW laser). Figure 8 presents the SERS intensity changes in MB molecules on the 3D stacked Ag NWs (Figure 8a) and composite substrate (Figure 8b) as the cycle time was increased with laser illumination at 638 nm and 20 mW. Prior to the photodegradation process, a high-intensity characteristic MB Raman spectrum was obtained from the 3D stacked Ag NWs (black line in Figure 8a) using 10 mW 638 nm laser illumination over 0.01 s. Because the 3D stacked Ag NWs had a high density of hot spots (crossed regions between Ag NWs) for use in SERS [18,21,27,44], we obtained highly enhanced MB SERS spectra from the 3D Ag NWs. The SERS intensity of MB decreased significantly after the first photodegradation process (10 s illumination at 20 mW, red line in Figure 8a). The SERS intensity remained unchanged during additional photodegradation cycles. We performed these cyclic photodegradation and SERS measurements using a 3D composite substrate (Figure 8b). Interestingly, the initial SERS intensity of the MB obtained from the 3D composite substrate increased beyond 40%, compared to the spectrum obtained from the 3D stacked Ag NWs. This increase was attributed to the compact integration of 3D TiO_2 NPs on the crossed regions of the Ag NWs. Because 3D volumetric hot spots formed, more MB molecules could be deposited onto the TiO_2 NPs and Ag NWs within the 3D hot spot volume [43]. The MB SERS intensity decreased significantly after only one photodegradation process (10 s illumination at 20 mW, red line in Figure 8b). Plasmonic metal nanostructures strongly absorb the light, leading to significant local heating associated with the resonant plasmonic excitations [45,46]. This thermoplasmonic effect, therefore, partially enabled the plasmon-induced reactions, particularly in the cross-point regions of Ag NWs [47].

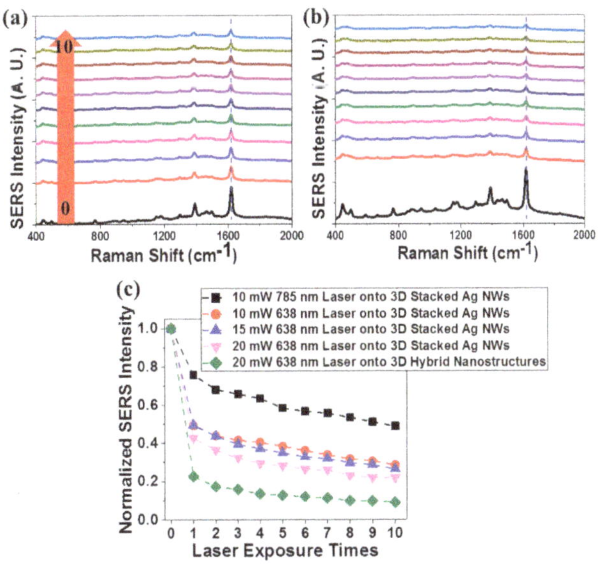

Figure 8. SERS intensity changes in the MB deposited onto (**a**) 3D stacked Ag NWs structures and (**b**) 3D hybrid (0.002 wt% TiO_2/Ag NWs) nanostructures during 638 nm laser exposure (10 s). The SERS measurements were performed by illuminating with a 638 nm laser for only 0.01 s to avoid photocatalytic effects. (**c**) The overall SERS intensity changes at 1620 cm^{-1} in MB molecules under different illumination conditions.

Figure 8c presents the overall SERS intensity changes at 1620 cm^{-1} of MB molecules under different illumination conditions. Figure 8c reveals several important characteristics of the plasmon-enhanced photocatalysis reaction. First, the 3D stacked Ag NWs displayed higher photocatalytic activities at shorter illumination wavelengths, in agreement with simulation results of local field enhancements and the optical extinction spectra of 3D plasmonic Ag NWs. Second, for a given photocatalyst and incident wavelength of light, increasing the incident light intensity increased the photocatalytic activity [48]. These results indicated that a quantitative comparison of the photocatalytic activities of the different catalysts required fixed light irradiation conditions, including the light source and intensity. Third, the 3D composite structures exhibited superior photocatalytic activities over the 3D Ag NWs, in good agreement with the photocatalytic performances. Finally, the photodegradation process mainly occurred at specific sites. We clearly observed that the SERS intensities of the probe molecules dramatically decreased only during the first cycle, then remained unchanged during the additional nine cycles. This result was attributed to the fact that prior to the photodegradation process, MB molecules deposited onto the crossed regions mainly accounted for the high SERS intensity. MB molecules on the hot spots decomposed during the first photodegradation process, and the SERS intensity decreased significantly due to the absence of target molecules in the hot spot. MB molecules deposited onto the non-crossed regions did not decompose, even after ten cycles, and MB molecules in non-hot spot regions (such as Ag NWs at the non-crossed region) contributed to a low SERS intensity. That is, hot spots were cross-points in the 3D stacked Ag NWs and hybrid TiO_2/Ag NWs structures for the SERS and photodegradation process. The spectroscopic studies of plasmon-enhanced photocatalysis revealed that excess irradiation onto hot spot areas could trigger the photo-induced elimination of target molecules, reducing the sensitivity (or increasing the limit of detection) in highly sensitive plasmonic nanostructures during SERS measurements.

4. Conclusions

In this paper, we proposed the use of new composite nanomaterial into which TiO_2 NPs were compactly integrated into the 3D cross-points of vertically stacked Ag NWs for plasmon-enhanced photocatalysis. The composite nanomaterials improved the photocatalytic activities under UV-visible and visible illumination due to the synergistic effects of visible light absorption by the 3D Ag NWs and efficient charge separation at the interface between the Ag NWs and TiO_2 NPs. We performed theoretical simulations of the local field enhancements by the 3D stacked Ag NWs illuminated at different wavelengths, and we systematically investigated the Raman spectra during plasmon-enhanced photocatalysis. These results revealed that the organic dyes underwent photo-induced decomposition mainly at the cross-points of the 3D vertically stacked Ag NWs and 3D hybrid nanostructures. As the photocatalytic activities were highly localized at specific regions of the 3D nanostructures, it is important to enhance the mass transport of reagents to the hot spot regions to boost the photocatalytic performance.

Supplementary Materials: The following are available online at http://www.mdpi.com/2079-4991/9/3/468/s1, Figure S1: SEM images of 3D hybrid nanostructures upon application of a 0.01 wt% TiO_2 NPs solution onto 3D stacked Ag NWs. Small TiO_2 NPs covered entire pores of the 3D stacked Ag NW substrate; Figure S2: Extinction spectra of different nanomaterials. The black and green dashed lines indicate the interband transitions of Ag and the band gap of TiO_2, respectively; Figure S3: Simulation setup of the 3D stacked Ag NWs. The inset shows a region of interest for a simulation of the cross-points between Ag NWs. The TEM image clearly shows that the Ag NWs were pentagonal in shape; Figure S4: Local electric field intensities at the junctions of crossed Ag NWs. The FDTD method was used in the calculations, along with 450 nm incident. The maximum field enhancement was obtained by searching the maximum values in the domain of the figure. The maximum values were located at the left and right bottom corners of the central pentagon, as shown in (a) and (b). The average field enhancement along the central Ag NW surface was extracted, as indicated by the green lines (0.5 nm away from the Ag NW surface to avoid the staircase effect in the simulation); Figure S5: Absorbance changes in the MB solutions after 10 min illumination. 10 mL of 0.05 mM MB aqueous solution were mixed with 4 mL of 0.01 wt% TiO_2 NPs aqueous solution.

Author Contributions: S.-G.P. designed and directed the research; V.T.N.L and H.S.J. fabricated the samples and performed the photocatalytic experiments; X.F.X., V.G., and S.A.M performed the plasmonic simulation; S.-G.P.,

D.-H. K and Y.-I.L. revised and discussed the results; S.-G.P. wrote the manuscript. All the authors read and commented on the manuscript.

Funding: This research was supported by the Fundamental Research Program (PNK 6070) of the Korean Institute of Materials Science (KIMS). This work was supported by the "Ministry of Trade, Industry and Energy". (Grant N0002310).

Acknowledgments: S. A. Maier acknowledges ONR Global (N00014-16-1-2288), the EPSRSC Reactive Plasmonics Programme (EP/M013812/1), the Lee-Lucas Chair in Physics, and the Bavarian Solar Technologies Go Hybrid (SOLTECH) Programme. X. Xiao is supported by Lee Family Scholars.

Conflicts of Interest: The authors declare no conflict of interest.

References

1. Seh, Z.W.; Liu, S.; Zhang, S.Y.; Bharathi, M.S.; Ramanarayan, H.; Low, M.; Shah, K.W.; Zhang, Y.W.; Han, M.Y. Anisotropic growth of titania onto various gold nanostructures: Synthesis, theoretical understanding, and optimization for catalysis. *Angew. Chem. Int. Ed.* **2011**, *50*, 10140–10143. [CrossRef] [PubMed]
2. Cushing, S.K.; Li, J.; Meng, F.; Senty, T.R.; Suri, S.; Zhi, M.; Li, M.; Bristow, A.D.; Wu, N. Photocatalytic activity enhanced by plasmonic resonant energy transfer from metal to semiconductor. *J. Am. Chem. Soc.* **2012**, *134*, 15033–15041. [CrossRef] [PubMed]
3. Zhang, X.; Zhu, Y.; Yang, X.; Wang, S.; Shen, J.; Lin, B.; Li, C. Enhanced visible light photocatalytic activity of interlayer-isolated triplex Ag@SiO_2@TiO_2 core-shell nanoparticles. *Nanoscale* **2013**, *5*, 3359–3366. [CrossRef] [PubMed]
4. Meng, X.W.; Li, B.; Ren, X.L.; Tan, L.F.; Huang, Z.B.; Tang, F.Q. One-pot gradient solvothermal synthesis of Au-Fe_3O_4 hybrid nanoparticles for magnetically recyclable catalytic applications. *J. Mater. Chem. A* **2013**, *1*, 10513–10517. [CrossRef]
5. Silva, C.G.; Juárez, R.; Marino, T.; Molinari, H.; García, H. Influence of excitation wavelength (uv or visible light) on the photocatalytic activity of titania containing gold nanoparticles for the generation of hydrogen or oxygen from water. *J. Am. Chem. Soc.* **2011**, *133*, 595–602. [CrossRef]
6. Lee, J.; Mubeen, S.; Ji, X.; Stucky, G.D.; Moskovits, M. Plasmonic photoanodes for solar water splitting with visible light. *Nano Lett.* **2012**, *12*, 5014–5019. [CrossRef] [PubMed]
7. Mubeen, S.; Lee, J.; Singh, N.; Kramer, S.; Stucky, G.D.; Moskovits, M. An autonomous photosynthetic device in which all charge carriers derive from surface plasmons. *Nat. Nanotechnol.* **2013**, *8*, 247–251. [CrossRef] [PubMed]
8. An, C.; Wang, J.; Jiang, W.; Zhang, M.; Ming, X.; Wang, S.; Zhang, Q. Strongly visible-light responsive plasmonic shaped Agx:Ag (x = cl, br) nanoparticles for reduction of CO_2 to methanol. *Nanoscale* **2012**, *4*, 5646–5650. [CrossRef]
9. Wang, C.; Ranasingha, O.; Natesakhawat, S.; Ohodnicki, P.R., Jr.; Andio, M.; Lewis, J.P.; Matranga, C. Visible light plasmonic heating of au-zno for the catalytic reduction of CO_2. *Nanoscale* **2013**, *5*, 6968–6974. [CrossRef]
10. Mankidy, B.D.; Joseph, B.; Gupta, V.K. Photo-conversion of CO_2 using titanium dioxide: Enhancements by plasmonic and co-catalytic nanoparticles. *Nanotechnology* **2013**, *24*, 405402. [CrossRef]
11. Zhang, Q.; Lima, D.Q.; Lee, I.; Zaera, F.; Chi, M.; Yin, Y. A highly active titanium dioxide based visible-light photocatalyst with nonmetal doping and plasmonic metal decoration. *Angew. Chem. Int. Ed.* **2011**, *50*, 7088–7092. [CrossRef] [PubMed]
12. Hou, W.; Cronin, S.B. A review of surface plasmon resonance-enhanced photocatalysis. *Adv. Funct. Mater.* **2013**, *23*, 1612–1619. [CrossRef]
13. Xiao, M.; Jiang, R.; Wang, F.; Fang, C.; Wang, J.; Yu, J.C. Plasmon-enhanced chemical reactions. *J. Mater. Chem. A* **2013**, *1*, 5790–5805. [CrossRef]
14. Wang, P.; Huang, B.; Dai, Y.; Whangbo, M.H. Plasmonic photocatalysts: Harvesting visible light with noble metal nanoparticles. *Phys. Chem. Chem. Phys.* **2012**, *14*, 9813–9825. [CrossRef] [PubMed]
15. Halasi, G.; Kecskeméti, A.; Solymosi, F. Photocatalytic reduction of no with ethanol on Ag/TiO_2. *Catal. Lett.* **2010**, *135*, 16–20. [CrossRef]
16. Wu, F.; Hu, X.; Fan, J.; Liu, E.; Sun, T.; Kang, L.; Hou, W.; Zhu, C.; Liu, H. Photocatalytic activity of Ag/TiO_2 nanotube arrays enhanced by surface plasmon resonance and application in hydrogen evolution by water splitting. *Plasmonics* **2013**, *8*, 501–508. [CrossRef]

17. Jiang, R.; Li, B.; Fang, C.; Wang, J. Metal/semiconductor hybrid nanostructures for plasmon-enhanced applications. *Adv. Mater.* **2014**, *26*, 5274–5309. [CrossRef] [PubMed]
18. Lee, M.; Mun, C.; Kim, D.-H.; Chang, S.-C.; Park, S.-G. Analyte-concentrating 3d hybrid plasmonic nanostructures for use in highly sensitive chemical sensors. *RSC Advances* **2016**, *6*, 92120–92126. [CrossRef]
19. Johnson, P.B.; Christy, R.W. Optical constants of the noble metals. *Phys. Rev. B* **1972**, *6*, 4370–4379. [CrossRef]
20. Park, S.G.; Mun, C.; Lee, M.; Jeon, T.Y.; Shim, H.S.; Lee, Y.J.; Kwon, J.D.; Kim, C.S.; Kim, D.H. 3d hybrid plasmonic nanomaterials for highly efficient optical absorbers and sensors. *Adv. Mater.* **2015**, *27*, 4290–4295. [CrossRef]
21. Koh, E.H.; Mun, C.; Kim, C.; Park, S.G.; Choi, E.J.; Kim, S.H.; Dang, J.; Choo, J.; Oh, J.W.; Kim, D.H.; et al. M13 bacteriophage/silver nanowire surface-enhanced raman scattering sensor for sensitive and selective pesticide detection. *ACS Appl. Mater. Interfaces* **2018**, *10*, 10388–10397. [CrossRef]
22. Molinari, R.; Palmisano, L.; Drioli, E.; Schiavello, M. Studies on various reactor configurations for coupling photocatalysis and membrane processes in water purification. *J. Membr. Sci.* **2002**, *206*, 399–415. [CrossRef]
23. Molinari, R.; Pirillo, F.; Loddo, V.; Palmisano, L. Heterogeneous photocatalytic degradation of pharmaceuticals in water by using polycrystalline TiO_2 and a nanofiltration membrane reactor. *Catal. Today* **2006**, *118*, 205–213. [CrossRef]
24. Chong, M.N.; Jin, B.; Chow, C.W.K.; Saint, C. Recent developments in photocatalytic water treatment technology: A review. *Water Res.* **2010**, *44*, 2997–3027. [CrossRef] [PubMed]
25. Lee, S.-A.; Choo, K.-H.; Lee, C.-H.; Lee, H.-I.; Hyeon, T.; Choi, W.; Kwon, H.-H. Use of ultrafiltration membranes for the separation of TiO_2 photocatalysts in drinking water treatment. *Ind. Eng. Chem. Res.* **2001**, *40*, 1712–1719. [CrossRef]
26. Bavykin, D.V.; Friedrich, J.M.; Walsh, F.C. Protonated titanates and TiO_2 nanostructured materials: Synthesis, properties, and applications. *Adv. Mater.* **2006**, *18*, 2807–2824. [CrossRef]
27. Jeon, T.Y.; Park, S.-G.; Lee, S.Y.; Jeon, H.C.; Yang, S.M. Shape control of Ag nanostructures for practical SERS substrates. *ACS Appl. Mater. Interfaces* **2013**, *5*, 243–248. [CrossRef] [PubMed]
28. Jeon, T.Y.; Kim, D.J.; Park, S.G.; Kim, S.H.; Kim, D.H. Nanostructured plasmonic substrates for use as sers sensors. *Nano Converg.* **2016**, *3*, 18. [CrossRef]
29. Jeong, J.W.; Arnob, M.M.; Baek, K.M.; Lee, S.Y.; Shih, W.C.; Jung, Y.S. 3d cross-point plasmonic nanoarchitectures containing dense and regular hot spots for surface-enhanced raman spectroscopy analysis. *Adv. Mater.* **2016**, *28*, 8695–8704. [CrossRef]
30. Jeong, J.W.; Yang, S.R.; Hur, Y.H.; Kim, S.W.; Baek, K.M.; Yim, S.; Jang, H.I.; Park, J.H.; Lee, S.Y.; Park, C.O.; et al. High-resolution nanotransfer printing applicable to diverse surfaces via interface-targeted adhesion switching. *Nat. Commun.* **2014**, *5*, 5387. [CrossRef]
31. Garnett, E.C.; Cai, W.; Cha, J.J.; Mahmood, F.; Connor, S.T.; Greyson Christoforo, M.; Cui, Y.; McGehee, M.D.; Brongersma, M.L. Self-limited plasmonic welding of silver nanowire junctions. *Nat. Mater.* **2012**, *11*, 241. [CrossRef]
32. Sau, T.K.; Rogach, A.L.; Jackel, F.; Klar, T.A.; Feldmann, J. Properties and applications of colloidal nonspherical noble metal nanoparticles. *Adv. Mater.* **2010**, *22*, 1805–1825. [CrossRef] [PubMed]
33. Jung, H.H.; Kwon, J.-D.; Lee, S.; Su Kim, C.; Nam, K.-S.; Jeong, Y.; Chung, K.-B.; Yoon Ryu, S.; Ocak, T.; Eray, A.; et al. Doping-free silicon thin film solar cells using a vanadium pentoxide window layer and a lif/al back electrode. *Appl. Phys. Lett.* **2013**, *103*, 073903. [CrossRef]
34. Mondal, C.; Pal, J.; Ganguly, M.; Sinha, A.K.; Jana, J.; Pal, T. A one pot synthesis of Au–ZnO nanocomposites for plasmon-enhanced sunlight driven photocatalytic activity. *New J. Chem.* **2014**, *38*, 2999–3005. [CrossRef]
35. Linic, S.; Aslam, U.; Boerigter, C.; Morabito, M. Photochemical transformations on plasmonic metal nanoparticles. *Nat. Mater.* **2015**, *14*, 567–576. [CrossRef] [PubMed]
36. Brongersma, M.L.; Halas, N.J.; Nordlander, P. Plasmon-induced hot carrier science and technology. *Nat. Nanotechnol.* **2015**, *10*, 25–34. [CrossRef]
37. Aslam, U.; Chavez, S.; Linic, S. Controlling energy flow in multimetallic nanostructures for plasmonic catalysis. *Nat. Nanotechnol.* **2017**, *12*, 1000–1005. [CrossRef] [PubMed]
38. Robatjazi, H.; Bahauddin, S.M.; Doiron, C.; Thomann, I. Direct plasmon-driven photoelectrocatalysis. *Nano Lett.* **2015**, *15*, 6155–6161. [CrossRef] [PubMed]

39. Christopher, P.; Xin, H.; Marimuthu, A.; Linic, S. Singular characteristics and unique chemical bond activation mechanisms of photocatalytic reactions on plasmonic nanostructures. *Nat. Mater.* **2012**, *11*, 1044–1050. [CrossRef] [PubMed]
40. Linic, S.; Christopher, P.; Xin, H.; Marimuthu, A. Catalytic and photocatalytic transformations on metal nanoparticles with targeted geometric and plasmonic properties. *Acc. Chem. Res.* **2013**, *46*, 1890–1899. [CrossRef] [PubMed]
41. Li, X.; Chen, G.; Yang, L.; Jin, Z.; Liu, J. Multifunctional Au-coated TiO_2 nanotube arrays as recyclable sers substrates for multifold organic pollutants detection. *Adv. Funct. Mater.* **2010**, *20*, 2815–2824. [CrossRef]
42. Park, S.-G.; Jeon, T.Y.; Jeon, H.C.; Kwon, J.-D.; Mun, C.; Lee, M.; Cho, B.; Kim, C.S.; Song, M.; Kim, D.-H. Fabrication of Au-decorated 3d ZnO nanostructures as recyclable sers substrates. *IEEE Sens. J.* **2016**, *16*, 3382–3386. [CrossRef]
43. Park, S.-G.; Mun, C.; Xiao, X.; Braun, A.; Kim, S.; Giannini, V.; Maier, S.A.; Kim, D.-H. Surface energy-controlled sers substrates for molecular concentration at plasmonic nanogaps. *Adv. Funct. Mater.* **2017**, *27*, 1703376. [CrossRef]
44. Chen, M.; Phang, I.Y.; Lee, M.R.; Yang, J.K.; Ling, X.Y. Layer-by-layer assembly of Ag nanowires into 3d woodpile-like structures to achieve high density "hot spots" for surface-enhanced raman scattering. *Langmuir* **2013**, *29*, 7061–7069. [CrossRef]
45. Politano, A.; Di Profio, G.; Fontananova, E.; Sanna, V.; Cupolillo, A.; Curcio, E. Overcoming temperature polarization in membrane distillation by thermoplasmonic effects activated by ag nanofillers in polymeric membranes. *Desalination* **2019**, *451*, 192–199. [CrossRef]
46. Herzog, J.B.; Knight, M.W.; Natelson, D. Thermoplasmonics: Quantifying plasmonic heating in single nanowires. *Nano Lett.* **2014**, *14*, 499–503. [CrossRef]
47. Politano, A.; Cupolillo, A.; Di Profio, G.; Arafat, H.A.; Chiarello, G.; Curcio, E. When plasmonics meets membrane technology. *J. Phys.: Condens. Matter* **2016**, *28*, 363003. [CrossRef]
48. Kale, M.J.; Avanesian, T.; Christopher, P. Direct photocatalysis by plasmonic nanostructures. *ACS Catal.* **2013**, *4*, 116–128. [CrossRef]

© 2019 by the authors. Licensee MDPI, Basel, Switzerland. This article is an open access article distributed under the terms and conditions of the Creative Commons Attribution (CC BY) license (http://creativecommons.org/licenses/by/4.0/).

nanomaterials

MDPI

Article

Efficient Copper Removal from an Aqueous Anvironment using a Novel and Hybrid Nanoadsorbent Based on Derived-Polyethyleneimine Linked to Silica Magnetic Nanocomposites

Olivija Plohl [1,*], Matjaž Finšgar [2], Sašo Gyergyek [2,3], Urban Ajdnik [1], Irena Ban [2] and Lidija Fras Zemljič [1]

1 Laboratory for Characterization and Processing of Polymers, University of Maribor, Faculty of Mechanical Engineering, Smetanova 17, 2000 Maribor, Slovenia; urban.ajdnik@um.si (U.A.); lidija.fras@um.si (L.F.Z.)

2 Faculty of Chemistry and Chemical Engineering, University of Maribor, Smetanova 17, 2000 Maribor, Slovenia; matjaz.finsgar@um.si (M.F.); saso.gyergyek@ijs.si (S.G.); irena.ban@um.si (I.B.)

3 Department for Materials` Synthesis, Jožef Stefan Institute, Jamova 39, 1000 Ljubljana, Slovenia

* Correspondence: olivija.plohl@um.si; Tel.: +386-2220-7715

Received: 10 January 2019; Accepted: 30 January 2019; Published: 6 February 2019

Abstract: Due to the extreme rise of sludge pollution with heavy metals (e.g. copper), the options for its disposal or treatment are decreasing. On the contrary, properly heavy metal-cleaned sludge can be used as an alternative sustainable energy and agriculture source. The aim of this study was to develop a novel nanoadsorbent, based on irreversibly linked amino-rich polymer onto previously silica-coated magnetic nanoparticles (MNPs) that can be applied efficiently for metal removal. MNPs were coated uniformly by 3 nm thick silica layer (core-shell structure), and were additionally modified with systematic covalent attachment of derived branched polyethyleneimine (bPEI). The formed structure of synthesized MNPs composite was confirmed with several analytical techniques. Importantly, nanoadsorbents exhibit high density of chelating amino groups and large magnetic force for easier separation. The importance of introduced bPEI, effect of pH, initial heavy metal concentration onto copper uptake efficiency and, further, nanoadsorbent regeneration, were studied and explained in detail. The adsorption isotherm was well fitted with Langmuir model, and the maximum adsorption capacity was shown to be 143 $mg{\cdot}g^{-1}$ for Cu^{2+}. The reusability and superior properties of silica-coated MNPs functionalized with derived-bPEI for copper adsorption underlie its potential for the removal application from heavy metals contaminated sludge

Keywords: environmental nanotechnology; nanoparticle characterization; surface analysis; magnetic polymer nanosorbents; heavy metal reduction; hybrid nanocomposites

1. Introduction

During recent decades, the upgrade of municipal wastewater treatment plants has led to increased sludge production. Consequently, sludge quantities continue to increase exponentially, whereas disposal options are decreasing due to their contamination with inorganic pollutants, such as Heavy Metals (HMs), by obeying strict national legislation. For instance, in the EU, it is estimated that sludge production will reach approximately 13 million tons of dried matter by 2020. On the contrary, moderately HM-cleaned sewage sludge can be used beneficially on lands as a fertilizer or as a soil conditioner, which can be advantageous by providing an economical, renewable disposal alternative, together with the additional nutritious supplement for plants [1–5]. Although the contamination with specific HMs is dependent on the location and type of the produced sludge, the major HM contaminant in significantly higher amount, as the legislation allows, is copper [1]. Nowadays, copper is used widely in various manufacturing industries (e.g., Electroplating), which caused its huge accumulation in the sewage sludge [6]. Besides, copper can bio-accumulate in plants, and can be deposited in the human body via the food chain. Despite the known necessity of the latter as an indispensable element for plants and humans, an excessive amount of copper can result in severe health problems [7]. Following this issue, there is an urgent need to remove the excess copper from sewage sludge to provide recycling (zero waste concept) and economical disposal of sludge with high nutrients values (i.e., phosphorus, nitrogen, organic matter) for agricultural land applications.

To date, several conventional methods for HMs' removal from contaminated wastewater have been introduced, such as chemical precipitation, coagulation, ultrafiltration, and adsorption [8–10]. Among them, a lot of attention has been paid to the adsorption process, as it allows higher efficiency, lower operational costs accompanied with lower energy use, process reversibility and flexibility in adsorbent design [11,12]. Most of the newly established adsorbents were based on active carbon, graphene oxide, mesoporous silica, or zeolites [11]. However, their major drawback is their complicated removal after the adsorption process (filtration, centrifugation), correlated with high costs, that are economically unattractive. Bearing this in mind, magnetic adsorbents, especially those in the nanoscale size range, can seem to be a great solution. Adsorbents' magnetic response can serve as a simple and costless separation process, whereas nano-size and large specific surface area enables more active adsorption sites for HM removal. Such magnetic adsorbents based on magnetic iron-oxide (MNPs) are environmentally friendly, costless, and can be produced on a large scale [8,13]. Despite their advantages, bare/unmodified MNPs seem to be less efficient in the adsorption process, as the bare MNPs' surface is relatively inert [14]. Therefore, they may be coated with a thick silica layer that improves the density of the –OH groups significantly, and can be further employed for chemical linkage with additional substrates [15]. Moreover, silica porosity increases surface area, it is chemically stable, costless, and shields the bare MNPs from dissolution in acidic aqueous media [16]. However, the thickness of the silica layer should be limited, because it dilutes the saturation magnetization [15].

The amino-(bio)polymers with nitrogen-donor functional groups can be used for chelation with HMs [17]. Such polymer, with amino-rich functional groups, is branched polyethyleneimine (bPEI), consisting of primary, secondary and ternary amines accompanied by highly hyper-branched structure [18]. Because of its solubility in an aqueous environment, it is often immobilized on the particle/support. The combination of the latter has been shown to be not only a biocompatible material in biomedical applications [19], but also in HM removal applications [20–22], together with anionic dyes and disinfection [23–25]. Alternatively, due to the high density of functional groups (i.e., primary amines), bPEI can be crosslinked to an organic functional reactive group, such as the epoxy group. Covalent coupling to polymeric particles possessing primary amine functionalities at the polymers' backbone through reaction with the epoxy-organosilanes containing groups, results in secondary amine and ring-opening reaction facilitated under alkaline conditions. One of the very useful organosilane modification agents is 3-glycidoxypropyltrimethoxysilane (GOPTS). GOPTS carries reactive epoxy groups on one side of the molecule, and a silane group on the other side. Inorganic surfaces coated with an amorphous silica layer can be used for covalent linkage with amine-containing ligands at specific pH of the reaction. In this context, GOPTS can be used to link the inorganic silica (terminated with abundant –OH groups) with the ligand, having a primary amine group [26].

It is generally known that metal adsorption efficiency is highly dependent on good magnetic response, large surface-to-volume ratio, many active bonding sites, and the irreversibly linked environmentally-friendly adsorbent on a magnetic carrier [8,13]. Taking this into consideration, the MNPs coated with a thick silica layer and chemically modified with bPEI, could represent an ideal adsorbent. Recent studies attached bPEI onto MNPs via electrostatic and/or physical interactions [21,27–29], causing unstable coatings. Furthermore, hazardous chemicals were used [30], or the process included several synthesis steps to achieve the desired product, being rather unattractive economically [20,24,31]. Regarding bPEI-based magnetic adsorbent, studies were focused mainly on the removal of chromium [31–33] or lead [22,28,30], whereas such kinds of polyethyleneimine-derived silica-coated MNPs for copper removal were not found in the relevant literature. Although a similar study for linkage of MNPs@SiO_2 with bPEI was recently reported [34], their synthesis concept and methodology differ significantly from our study. Moreover, the designed micro-scale adsorbent was studied for uranium removal applications.

Surprisingly, to the best of our knowledge, no detailed and systematic study regarding the crosslinking of GOPTS with bPEI and its further silanization directly onto core-shell silica-coated MNPs for Cu^{2+} removal has been reported yet. There are several advantages of the introduced amino-based nanoadsorbents with respect to current state-of-the-art adsorbents: (i) Environmentally-friendly adsorbent, synthesized easily with the concept of green chemistry; (ii) The combination of the amino-biopolymer together with silica-modified MNPs ensures more functional sites for covalent bonding, easy manoeuvrability and removal of the adsorbent with no need for additional operating costs; (iii) Strongly covalently attached GOPTS-modified bPEI onto silica-coated MNPs ensures a strongly, permanently bonded amino-polymer that does not desorb during the cleaning procedure, and, consequently, improves adsorption efficiency, (iv) Systematic and detailed characterization of the physicochemical properties in terms of surface chemistry, charge, morphology, ligand density, etc. and (v) Plenty of primary amine groups from the bPEI macromolecule backbone that can serve as an HM chelator. An important, but often overlooked aspect, is that the efficient adsorption ability, together with quick magnetic response, should also be considered.

This work represents a comprehensive study of chemically modified bPEI with an epoxy group, and its silanization onto core-shell MNPs@SiO_2 for copper removal application. The synthesis parameters (silica shell thickness, amount of derived-bPEI and pH) were studied in detail, and accompanied by extensive nanoadsorbent characterization. The adsorption capacity was studied with Cu^{2+} aqueous model solutions to define the optimal parameters for transfer into real applications. The results revealed superior MNPs@SiO_2@GOPTS-bPEI ability for Cu^{2+} removal due to the high density of amino groups. Moreover, adsorption studies were performed analogously for MNPs@SiO_2, emphasizing the importance of the introduced amino-biopolymer on removal efficiency. Additionally, the mechanism of copper adsorption was proposed, employing different surface analysis techniques.

2. Materials and Methods

2.1. Materials and Chemicals:

$FeSO_4{\cdot}7H_2O$ was purchased from Honeywell (Seelze, Germany). $Fe_2(SO_4)_3{\cdot}7H_2O$, HCl (≥37%), tetraethylorthosilicate (TEOS, ≥98%) and branched PEI with average Molecular Weight (MW) of 25,000 Da, copper(II) chloride (97%) and 3-glycidoxypropyltrimethoxysilane (GOPTS, 98%) were all purchased from Sigma-Aldrich (Taufkirchen, Germany). NH_4OH (25% aqueous solution), NaOH (>98%) and acetone (≥99.5%) were purchased from Honeywell (Seelze, Germany). Absolute EtOH (anhydrous) was obtained from CarloErba (Val de Reuil, France), and citric acid (≥99.5%, water free) from Roth (Karlsruhe, Germany). All chemicals were used as received, without any further purification. Ultrapure water (with a resistivity of 18.2 MΩ cm, obtained from Milli-Q, Millipore Corporation, MA, USA) was used throughout the experiments.

2.2. Preparation of γ-Fe_2O_3 (MNPs) and SiO_2-coated MNPs (MNPs@SiO_2)

Magnetic nanoparticles based on maghemite, abbreviated hereinafter as MNPs, were synthesized under air atmosphere with coprecipitation [35]. Fe^{2+} and Fe^{3+} aqueous solution (V = 500 mL) with the $n(Fe^{2+}):n(Fe^{3+})$ = 2.4:1 ratio was prepared, using iron sulphate reagents, salts and ultrapure water. In order to precipitate iron hydroxides, a diluted aqueous ammonia solution was added slowly to an iron salts solution at pH 3. For precipitation and subsequent formation of magnetic iron oxide NPs, 250 mL of ammonia solution (25%) was added to the above mixture and agitated additionally for 30 min. Synthesized MNPs were washed several times with the diluted ammonia solution and ultrapure water. Afterwards, the colloidal stable dispersion of MNPs was prepared using the adsorption of Citric Acid (CA) in accordance to Reference [35]. Approximately 1.2 g of as-synthesized bare MNPs were redispersed in 60 mL of ultrapure water, together with the addition of 5 mL (0.5 g·mL^{-1}) CA aqueous solution. Dispersion pH was raised to 5.2 with the diluted ammonia solution, and put under reflux for 1.5 h at 80 °C. Following refluxing, the pH of the cooled dispersion was set to ~10 with ammonia solution (25%). The coating of MNPs stabilized with CA with silica shell (MNPs@SiO_2) was applied in similar way as in Reference [15], by introducing some modifications to achieve a 3 nm thick silica layer. NH_4OH was added to the MNPs@CA dispersion (15 mg·mL^{-1}, pH = 10.6). The mixture was agitated for 15 min and added rapidly to the solution of EtOH and TEOS (10 mg·mL^{-1}). This was followed with pH settling to 10.6, using 25% NH_4OH. The coating procedure was left to proceed for 2 h under continuous stirring. The obtained core-shell MNPs@SiO_2 were cleaned of the excess reagents using absolute EtOH and ultrapure water.

2.3. Systematic Covalent Linkage of GOPTS and bPEI (GOPTS-bPEI) onto MNPs@SiO_2 (i.e., MNPs@SiO_2@GOPTS-bPEI)

In order to perform synthesis systematically, some calculations were performed to avoid the excess of reagents, that can influence the MNPs@SiO_2 stability significantly during the modification process and decrease the silanization efficiency. The theoretical specific surface area of MNPs@SiO_2 was estimated to be around 95 m^2.g^{-1}, and the calculation was done in accordance with Čampelj et al. [35]. The estimated surface concentration of the Si-OH groups is 5 per 1 nm^2 [16], where each silanol group will hydrolyse with the methoxy-group from GOPTS. On the other hand, it was predicted that one epoxy ring will react with one primary amino group from bPEI (Scheme 1). Based on these predictions, the reaction conditions for synthesizing the MNPs@SiO_2@GOPTS-bPEI were as follows. For functionalization of 100 mg of MNPs@SiO_2 with GOPTS-bPEI, bPEI was dissolved in deionized water (V = 20 mL, 0.2 wt.%, bPEI value added to MNPs@SiO_2 corresponded to five monomers bPEI per 1 nm^2 of MNPs@SiO_2). pH was adjusted to 10 with 0.1 M HCl. Simultaneously, GOPTS (5 molecules per 1 nm^2 of MNPs@SiO_2) was dissolved in absolute EtOH (2 wt.%). After preparation of both solutions, the solution of GOPTS was added slowly to the bPEI aqueous solution, with the pH remaining at 10. It is widely accepted that the amine nucleophiles react with epoxy functionalities at moderate alkaline area (at least pH = 9) [26]. The mixture was left to stir for 15 min, and resulted in a clear solution with the absence of aggregates. The formed chemically coupled GOPTS-bPEI was then added to 0.4 wt.% MNPs@SiO_2 aqueous dispersion at pH = 10, where repulsive negative forces among the MNPs@SiO_2 should be strong enough to enable stable dispersion during the functionalization. Silanization reaction was left to proceed for 3 h under reflux at 60 °C. The proposed chemical coupling mechanism is shown schematically in Scheme 1. After chemical linkage, the dispersion was separated with a magnet and washed with acidic ultrapure water (pH = 4, adjusted with 0.1 M HCl) several times.

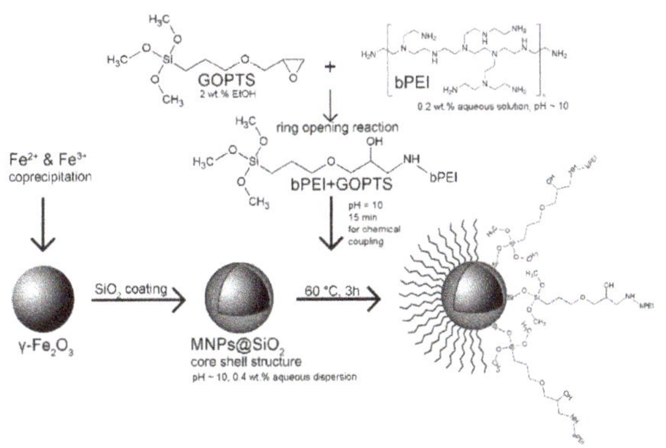

Scheme 1. Proposed chemical coupling mechanism for MNPs@SiO_2@GOPTS-bPEI.

2.4. Characterization of the Nanoadsorbent

The crystal structure and purity of as-synthesized bare MNPs was verified with X-Ray powder Diffraction (XRD) using a D-5005 diffractometer Bruker Siemens with CuKα radiation, $\lambda_{CuK\alpha}$ = 1.5406 Å. Parameters of the XRD measurements were 2θ from 30 to 70°, and the scan rate was 0.385° min^{-1}. Morphology, and nanoscale particle size of the bare and core-shell structure of coated MNPs were observed with Transmission Electron Microscopy (TEM, Jeol JEM-2100), operating at the acceleration voltage of 200 kV. For TEM investigations, dispersions were deposited onto a copper grid with transparent carbon film. Specific surface area, as well as equivalent diameter of particles, were estimated and calculated from the TEM images using Gatan Digital Micrograph Software. The Scanning Electron Microscope (Carl Zeiss FE-SEM SUPRA 35 VP) was used for the observation of the nanoadsorbent morphology. The nanoadsorbent was dried, put onto a double-sidde adhesive conductive carbon tape, placed onto an aluminium sample holder, and observed at an acceleration voltage of 1 keV, 30 µm-sized aperture, 5.5 mm working distance, and at 100,000× magnification. The surface modification steps were inspected with infrared spectroscopy (FTIR, PerkinElmer Spectrum GX spectrometer equipped with a diamond crystal Attenuated Total Reflection (ATR) attachment). For the latter, samples were dried, placed on a diamond crystal, and pressed into the thick film. Background, as well as all spectra, were reported as an average of 32 scans, that were recorded from 400–4000 cm^{-1} with a resolution of 2 cm^{-1} at room temperature. The weight fraction of mass loss was determined via thermogravimetric analysis (TGA/SDTA 851 Mettler Toledo). Dry samples were placed into 60 µL alumina crucibles and heated from 25 °C to 600 °C with a rate of 10 K·min^{-1} in an O_2 atmosphere. Zeta Potential (ZP) and hydrodynamic diameter, with corresponding number-sized distribution in aqueous dispersions, were monitored with the electrokinetic measurements via Dynamic Light Scattering by the electrophoresis experiments of the samples (DLS ZetaSizer Nano ZS, Malvern Instruments Ltd). For ZP measurements in aqueous dispersions, disposable folded capillary cell having electrodes was filled up and placed into a sample holder. Similarly, for the hydrodynamic diameter, the disposable cuvette was filled up (to 1 cm) with the aqueous dispersion. Particular pH measurement was adjusted with 0.1 M HCl or 0.1 M NaOH in 1 mM NaCl background at a constant temperature of 25 °C. At the end, the measurement data were collected with the usage of the manufacturer's software. pH-dependent potentiometric titrations were performed for the total charge determination, as well as for pK values' determination of the pure bPEI solution. The latter was carried out in a forward (acidic to alkaline) and backward direction (alkaline to acidic) at $2.5 < pH < 11.0$, using 0.1 M HCl and 0.1 M KOH aqueous solutions as titrants. A twin burette instrument (Mettler T-70) was equipped with a pH electrode (Mettler Toledo InLab Reach 225). All solutions for potentiometric titrations were prepared with ultrapure water having a very low carbonate content ($<10^{-6}$ M), and the ionic strength adjusted to 0.1 M with 3 M KCl. The blank titrations were performed under the same conditions as depicted above. A detailed description regarding the charge calculations and *p*K value determination can be found in Reference [36]. Room-temperature magnetization curves of samples were measured with a Vibrating-Sample Magnetometer (VSM, Lake Shore 7307). X-ray Photoelectron Spectroscopy (XPS) measurements were performed using a PHI-TFA 5600 XPS spectrometer from Physical Electronics Inc., which was equipped with a monochromatic Al X-ray excitation source (photon energy of 1486.6 eV) and a hemispherical electron analyzer. Spectra were acquired at 45° take-off angle. The radius of the analyzed area was 0.4 mm. An additional electron gun was used to compensate for the possible charging effect of the samples. Survey and high-resolution C 1s, N 1s, Si 2p and Fe 2p spectra were measured. The analysis of XPS spectra was performed with Multipak 8.1c software. A C 1s high-resolution spectrum was employed at a binding energy of 284.8 eV to correct the binding energy scale. The surface area and the total pore volume were determined with the adsorption–desorption of N_2 at 77 K, and analyzed further employing a BET analyzer (Micromeritics TriStar II Surface Area and Porosity), whereas results were subjected and analyzed with the TriStar II 3020 version 3.02 software package. To assure pure nanocomposites and the removal of any adsorbed contaminants, samples were degassed prior to surface area analysis.

2.5. Cu^{2+} Adsorption Experiments

A Cu^{2+} stock solution (5 g·L^{-1}) was prepared from $CuCl_2$. It is known that Cu^{2+} adsorbs [21] onto bPEI selectively, while the Cl^- was chosen as an anion, because the importance of an appropriate anion with respect on removal efficiency [37] has already been shown. Solutions for further experiments were prepared by dilution of the stock solution to the defined concentration (5−150 mg·L^{-1}, i.e., initial Cu^{2+} concentrations were 5, 10, 25, 50, 75, 100, and 150 mg·L^{-1}). The batch adsorption experiments were performed in 150 mL flasks under vigorous magnetic stirring. The 5 mg of nanoadsorbents were placed together with the 50 mL of Cu^{2+} solution in the flask at constant temperature (298 K), t = 1 h. The specific pH (3−6) was adjusted with the use of 0.1 M NaOH or HCl. After the adsorption process, the magnetic nanoadsorbent was removed immediately with the permanent magnet. Just to assure the elimination of any remaining nanoadsorbent, the supernatant was filtered through an MW 10 kDa filter using ultrafiltration. The supernatant was diluted with ultrapure water, and its quantity of residual Cu^{2+} was determined with Atomic Absorption Spectroscopy (AAS). AAS was employed for determination of the Cu^{2+} in supernatants after the adsorption process using a Perking Elmer 3110 spectrophotometer. Prior to analysis, the calibration curve of Cu^{2+} was prepared, followed by measurements of Cu^{2+} in solutions at a wavelength of 324.8 nm. Each measurement was done in triplicate, and the average results are reported. The Standard Deviation with this analytical technique was calculated to be 10% of the measured value. The adsorption capacity (q) and removal efficiency (R) were calculated with the following equations:

$$q\ (\mathrm{mg{\cdot}g^{-1}}) = (c_0 - c_t)V/m \tag{1}$$

$$R\ (\%) = (1 - c_t/c_0) \times 100 \tag{2}$$

q is the adsorption capacity (mg Cu^{2+} per g of the adsorbent), R is the removal efficiency, c_0 is the initial Cu^{2+} concentration in solution, and the c_t is the Cu^{2+} concentration in the supernatant after treatment, m is the mass of the nanoadsorbent (g), and V is the volume of the Cu^{2+} model solution.

Reusability cycles by means of adsorption–desorption experiments were performed with MNPs@SiO_2@GOPTS-bPEI by batch experiment, similar as already explained above. After finished adsorption, the magnetic nanoadsorbents were decanted onto the permanent magnet. For the following desorption studies, the nanoadsorbents with Cu were immersed in 10 mL of 0.1 M Na_2EDTA and left to agitate for 2 h. The Na_2EDTA was selected as an eluent, due to its known exceptional desorption capability [38]. Finally, MNPs@SiO_2@GOPTS-bPEI with desorbed Cu were again collected onto the permanent magnet, rinsed with ultrapure water, and applied further in the next reusability cycle.

3. Results and Discussion

3.1. Nanoadsorbent Characterization

The crystal structure of the magnetic part was verified with XRD analysis (Figure 1). The results revealed that the MNPs crystal structure was assigned to maghemite (Figure 1), as diffraction peaks corresponded to a standard reference card (JCPDS 72-0246, cubic space group Fd-3m), typical for a maghemite cubic spinel crystal structure. No other peaks were observed, indicating the phase purity of the synthesized bare MNPs. Additionally, broadening of the diffraction peaks shows the nanocrystallinity of the synthesized MNPs. In the continuation, regardless of the surface modification, the crystal structure remained unchanged.

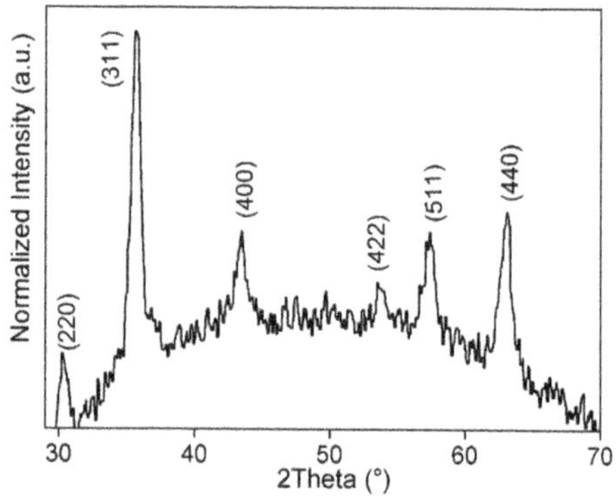

Figure 1. XRD pattern for synthesized bare magnetic nanoparticles (MNPs.)

The morphology and the shape of the silica-coated MNPs before and after functionalization with GOPTS-bPEI were investigated with TEM and SEM. Representative images are shown in Figure 2. It can be seen clearly that the MNPs (~13 nm) were of quasi-spherical shape, and were coated uniformly with around a 3 nm thick silica layer. The silica layer is surrounding the MNPs, which results in a core-shell structure (Figure 2a,b). The silica layer is amorphous, and it is seen in a higher contrast (gray) with respect to the core–magnetic part (dark). The optimal synthesis conditions reassured that almost all observed MNPs@SiO_2 had similar SiO_2-shell thickness with any observation of the homogeneously formed SiO_2 NPs. After the functionalization of MNPs@SiO_2 with the GOPTS-bPEI, no obvious influence on the morphology was observed with the additional silanization step (Figure 2b). We can conclude that grafting affects neither the morphology, nor the silica-shell thickness of the nanoadsorbent. A more clear morphology of the introduced MNPs@SiO_2@GOPTS-bPEI can be revealed from the SEM image shown in Figure 2c. The latter shows a quasi-spherical morphology that is in agreement with the TEM images, however, with the larger agglomeration degree that can be related to the difference of magnetic nanocomposite sample preparation for observation. Nevertheless, from Figure 2c, individual MNPs@SiO_2@GOPTS-bPEIs can also be distinguished, but at the lower magnification, as in the case of the TEM images.

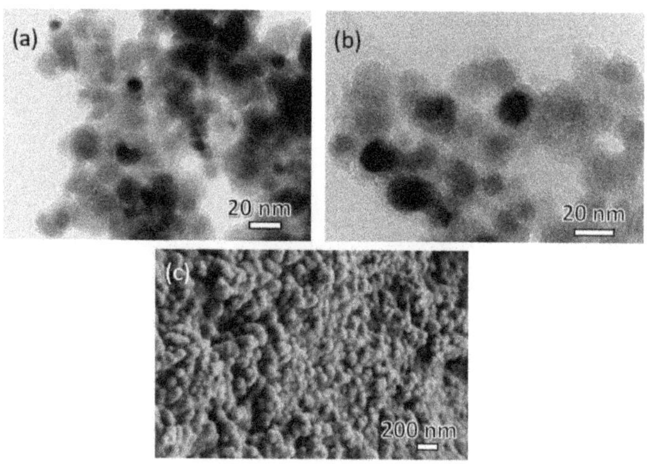

Figure 2. TEM image of MNPs@SiO_2 (**a**), TEM image of MNPs@SiO_2@GOPTS-bPEI (**b**), and representative SEM image of MNPs@SiO_2@GOPTS-bPEI (**c**). Note that darker spots in the TEM images correspond to the bare MNPs that overlap when deposited onto the copper grid.

The surface ATR-FTIR spectroscopy analysis confirmed the successful introduction of coatings (Figure 3). FTIR spectra of synthesized nanoadsorbent MNPs@SiO_2@GOPTS-bPEI with respect to the pure GOPTS and bPEI are shown in Figure 3a. The intensive band at 575 cm^{-1} originates from the magnetic part of the MNPs, and corresponds to the Fe–O vibrations of Fe_2O_3 [14]. The peak at around 1090 cm^{-1} corresponds to the Si–O–Si asymmetric stretching vibrations, representing the presence of SiO_2 coating in MNPs@SiO_2 [39]. Furthermore, the peaks at 2932 and 2845 cm^{-1}, ascribed to asymmetric stretching of CH_2 and symmetric stretching of CH groups, are evidently seen at MNPs@SiO_2@GOPTS-bPEI, confirming the presence of bPEI [40]. Moreover, the spectrum for pure bPEI shows a weak doublet at high frequencies, corresponding to primary amine [41]. After reaction of GOPTS with bPEI and its introduction onto MNPs@SiO_2, the latter is evidently not seen anymore. This can be correlated with the ring-opening process between the nucleophilic primary amine group of bPEI and epoxy group of GOPTS, which results in secondary amine formation (Scheme 1). Meanwhile, the epoxy bands vibrations located about 1270 and 830 cm^{-1} decreased significantly, indicating a ring-opening process and confirming our bonding mechanism between bPEI and GOPTS. To provide further information about efficient surface modification, the FTIR spectra of MNPs@SiO_2 was compared to the MNPs@SiO_2@GOPTS-bPEI (Figure 3b). The presence of peaks from the MNPs@SiO_2 core-shell nanocomposite, together with the newly introduced vibrational bands from GOPTS-bPEI, confirms the successful crosslinking. Specifically, the band at 627 cm^{-1} corresponding to the C-NH_2 vibrations, band at 1450 cm^{-1} attributed to C–N groups from bPEI and C–H stretching at 2932 and 2845 cm^{-1} [41], show the successful functionalization clearly.

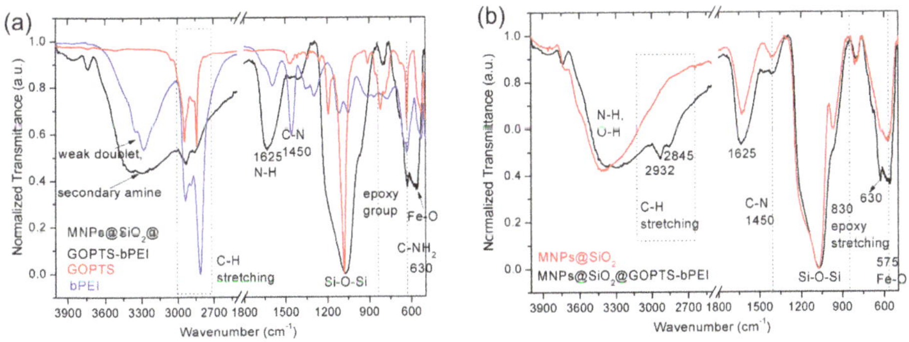

Figure 3. FTIR spectra of MNPs@SiO_2@GOPTS-bPEI with respect to pure bPEI and GOPTS (**a**). FTIR spectra of MNPs@SiO_2 and MNPs@SiO_2@GOPTS-bPEI (**b**).

The XPS technique is one of the most powerful surface sensitive techniques, and was employed in this study for the determination of the samples' surface atomic composition. The determined atomic concentrations for MNPs, MNPs@SiO_2, MNPs@SiO_2@GOPTS-bPEI are reported in Table 1. Corresponding survey and high-resolution spectra are shown in Figure 4.

The atomic concentration ratio of Fe:O for MNPs is in good agreement with the stoichiometric ratio of MNPs, and confirms Fe_2O_3–Fe (31.4 at.%) and O (53.2 at.%) (Table 1). A C signal was detected along with Fe and O. The latter is most likely a consequence of the adventitious carbonaceous species' adsorption (compounds that are present in the atmosphere) on the MNPs' surface during the preparation procedure [42]. Moreover, in agreement with the XRD results, the high-resolution Fe 2p spectra show an intensive peak at the binding energy of 710.9 eV (Figure 4b), confirming the Fe_2O_3 structure [43].

As expected, by further surface modification of MNPs with the amorphous silica layer, the Si signal was detected (MNPs@SiO_2, Figure 4a,c). Consequently, the Fe atomic concentration decreased significantly on the MNPs' surface (Table 1), due to the presence of the SiO_2 surface coating. The analyzed depth in the XPS studies is calculated as $3\lambda\sin\theta$ (an estimation where about 90% of the

XPS signal originates), where λ stands for the inelastic mean free path, and θ is the take-off angle (in the present case 45°). Therefore, as Fe was still detected, and due to the fact that the analyzed depth in this case is about 5 nm, the thickness of the SiO_2 is thinner than 5 nm. In fact, using TEM, it was determined to be around 3 nm (Figure 2a,b). High-resolution Si 2p spectra confirm that the source of Si originates from the SiO_2, as the peak is located at the binding energy of 103.5 eV (Figure 4c). Moreover, the SiO_2 surface coating on MNPs did not change the shape and peaks' position of the Fe 2p spectrum compared with that for MNPs, indicating that the surface coating with SiO_2 on the Fe_2O_3 structure did not modify the Fe_2O_3 (Figure 4b).

Followed by further functionalization of MNPs@SiO_2 with the chemically linked GOPTS−bPEI, the Fe signal is still detected (1.9 wt.%). However, the amount is lower than that for the MNPs@SiO_2, due to the even thicker surface coating regarding the additional layer. Additionally, the Si amount decreased to 14.6 at.%. On the contrary, with the MNPs@SiO_2 modification, the N signal was detected (9.1 at.%), which confirms the presence of bPEI. Moreover, the amount of the C also increased, as C atoms are the main constituents in the hyper-polymer branched structure. After the surface modification of the MNPs@SiO_2 with the GOPTS-bPEI, the Si 2p peak was transferred to the more negative binding energies (102.2 eV, Figure 4c) that can be attributed to the silanization process. The latter is in high agreement with our proposed bonding strategy (Scheme 1). The deconvoluted C 1s spectrum for MNPs@SiO_2@GOPTS-bPEI is shown in Figure 4d. The latter was fitted with two peaks, i.e., a feature at 284.8 eV corresponding to C–C/C–H bonds in the polymer backbone, and a feature at 286.7 eV that corresponds to C–O and/or C–N [44]. The presence of C–O species definitely originates from the cross-linked GOPTS-bPEI via a ring-opening reaction (Scheme 1), resulting in ether formation. On the other hand, C–N originates from the bounded bPEI. The deconvolution of the high-resolution N 1s spectrum for MNPs@SiO_2@GOPTS-bPEI showed that the major contribution corresponds to the C–N species in bPEI, while the deconvoluted high binding energy peak can be attributed to the small amount of protonated amino groups in the bPEI (Figure 4e) [43].

Table 1. Atomic concentrations determined with XPS for MNPs, core-shell MNPs@SiO_2, and GOPTS-bPEI modified MNPs@SiO_2.

	C	N	O	Si	Fe
MNPs	15.4		53.2		31.4
MNPs@SiO_2	7.4		65.4	24.4	3.3
MNPs@SiO_2@GOPTS-bPEI	29.5	9.4	43.4	16.0	1.9

Surface charge as a protonation/deprotonation ratio of functional groups is changing as a consequence of different modifications steps, and was correlated with the electrokinetic measurements. The latest results were supported with Dynamic Light Scattering. The comparison of Zeta Potential (ZP) values versus pH for MNPs, MNPs@SiO_2 and MNPs@SiO_2@GOPTS-bPEI are shown in Figure 5a. After coating of MNPs with an amorphous silica layer, SiO_2-coated MNPs expressed negative ZP values in almost the entire pH range (isoelectric point (IEP) at pH ~2.5) due to the presence of negatively charged silanol groups from the silica coating. With further GOPTS-bPEI functionalization to MNPs@SiO_2, the IEP shifted into an alkaline area (Figure 5a). At pH < 8.5, the amino groups were protonated, resulting in a positive charge [21], originating from the amino groups onto the bPEI-functionalized MNPs@SiO_2 surface. In contrast, at pH > 8.5, the amino groups at the MNPs@SiO_2@GOPTS-bPEI became deprotonated. As can be seen from Figure 5a, the shift of GOPTS-bPEI-functionalized MNPs@SiO_2 IEP to a higher pH value ($pH_{IEP} \approx 8.5$), which also confirms the successful covalent bonding of GOPTS-bPEI onto MNPs@SiO_2. Results are in high agreement with the pH-dependent potentiometric charge titration. It can be observed that the pure bPEI protonates/deprotonates in 3-pK fashion that lies between the plateaus at pH =3 and pH = 11 (Figure 5b). This is due to the contribution of the amine groups, which exhibit a highly positive charge (~15.7 $mmol \cdot g^{-1}$). As the bPEI is crosslinked with the GOPTS and functionalized onto MNPs@SiO_2 (Scheme 1), the $pH_{IEP} \approx 8.5$ is near to the pK_{a2} value of pure bPEI for the secondary amine, being around 7.8 (Figure 5b), which again proved the successful chemical coupling as proposed in Scheme 1. This phenomenon also supports the fact of the pure bPEI being in almost deprotonated state, expressing its high nucleophilicity when cross-linked to GOPTS at pH 10 (for details see the Experimental Section).

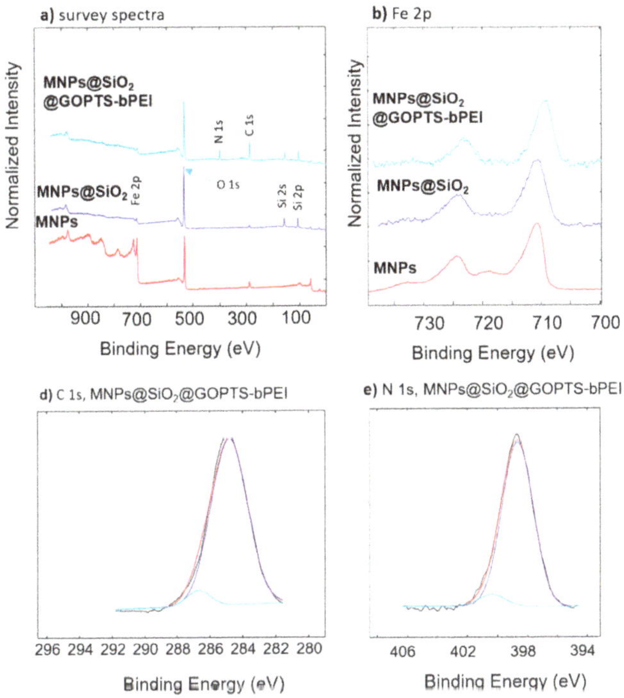

Figure 4. (**a**) Survey spectra and high-resolution, (**b**) Fe 2p and (**c**) Si 2p spectra. Deconvoluted high-resolution, (**d**) C 1s spectrum for MNPs@SiO_2@GOPTS-bPEI, and (**e**) N 1s spectrum MNPs@SiO_2@GOPTS-bPEI.

Dynamic Light Scattering was employed for estimation of the hydrodynamic diameter and size distribution when the nanoadsorbent was exposed to the aqueous environment. The latter were measured at pH = 6 (close to neutral range) for both aqueous dispersions (Figure 5c), where they exhibited the largest absolute ZP values according to the ZP measurements (Figure 5a). The hydrodynamic diameter of MNPs (~180 nm) is obviously larger in comparison to the MNPs@SiO_2 (~33 nm). This can be attributed to the high amount of anionic charged silanol groups from the MNPs@SiO_2, that provided high electrostatic repulsion and, thus, overcame the agglomeration. After the functionalization with GOPTS-bPEI, the hydrodynamic diameter increased (~92 nm), but the narrow size distribution is still observed. This fact correlated well with the highly positive nature of amine groups (i.e., secondary and ternary amine groups) being protonated and the hyper-branched structure of bPEI. Synergistically, both effects reassured the electrosteric repulsion. For the application of the nanoadsorbent, the nanosize and its distribution in aqueous media plays a crucial role in the achievement of better removal efficiency.

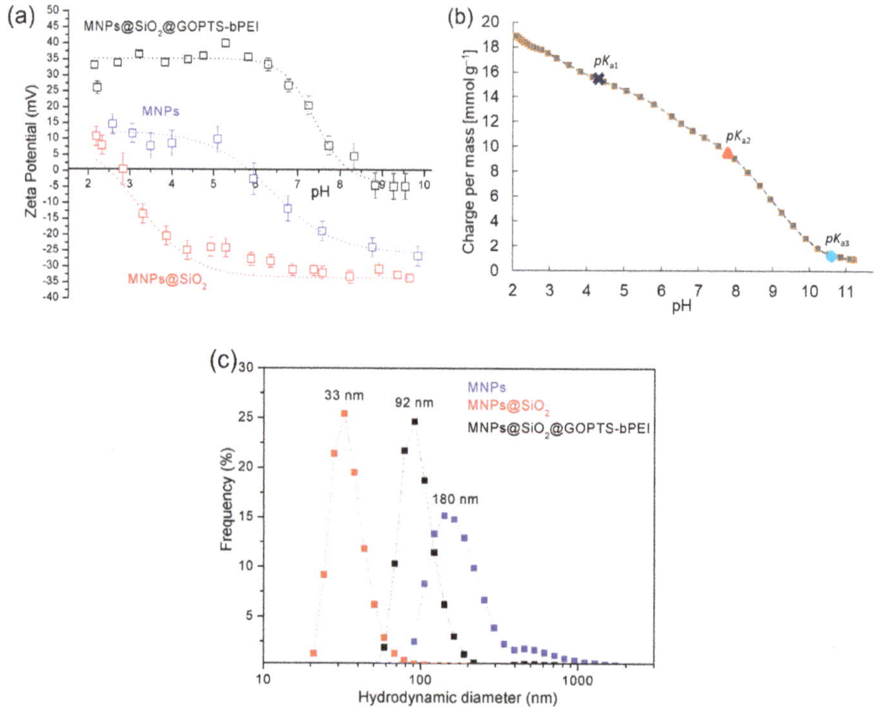

Figure 5. Zeta Potential measurement as a function of pH for SiO_2-coated MNPs and the same one functionalized with the GOPTS-bPEI (**a**); pH-depended potentiometric titration versus pH for pure hyper-branched polyethyleneimine (bPEI) (**b**) and number-sized distribution of hydrodynamic diameter at pH = 6 in aqueous media (**c**).

The synthesized magnetic nanocomposite was also characterized by means of the surface area and pore volume with the Brunauer–Elmett–Teller (BET) theory and Barett–Joyner–Halenda (BJH) method, respectively. It can be observed from Figure 6, that the adsorption–desorption isotherm graph of N_2 exhibits the typical IV type BET isotherm for MNPs@SiO_2@GOPTS-bPEI, and this phenomenon is attributed mainly to its mesoporous structure [45]. Quite a high specific area was measured for MNPs@SiO_2 (25.21 $m^2 \cdot g^{-1}$), and the latter is significantly increased with respect to the bare MNPs (for 86% larger surface area, Table 2) with introducing only a 3 nm thick layer. With additional surface modification using GOPTS-bPEI, the surface area increased to 38.33 $m^2 \cdot g^{-1}$ (52% increase) in comparison to the MNPs@SiO_2. Following the same trend, the total pore volume increased from 0.16 $cm^3 \cdot g^{-1}$ to 0.23 $cm^3 \cdot g^{-1}$ with additional functionalization (Table 2), and is in contrast to the reports [17,46], where they observed a decreased porous volume and surface area after functionalization with bPEI. This fact can be correlated to the modified MNPs@SiO_2 surface with GOPTS-bPEI that most probably causes the formation of additional intraparticle space, or pores originating from the derived polymer network structure. Indeed, the results indicate an even larger surface area and pore volume when modifying MNPs@SiO_2 surface by GOPTS-bPEI, showing an enlarged effective adsorption area towards successful HM removal.

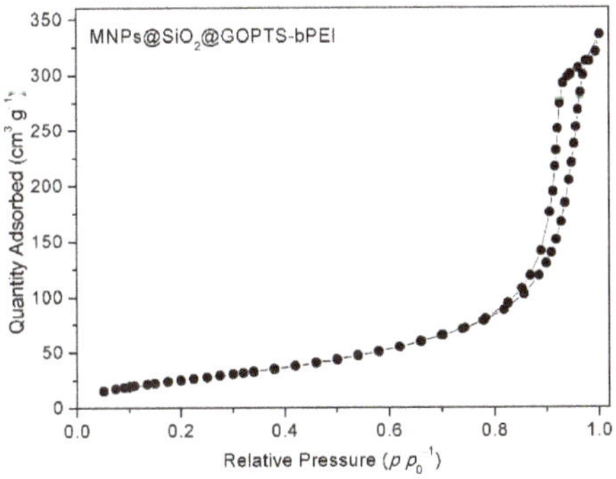

Figure 6. Graph of N_2 adsorption and desorption isotherm (77 K) as a function of relative pressure of MNPs@SiO_2@GOPTS-bPEI.

Table 2. Brunauer–Elmett–Teller (BET) surface area and pore volume for differently modified MNPs with respect to bare MNPs.

	BET Surface Area * ($m^2 \cdot g^{-1}$)	BJH Adsorption Pore Volume ($cm^3 \cdot g^{-1}$)
MNPs	13.56	/
MNPs@SiO_2	25.21	0.16
MNPs@SiO_2@GOPTS-bPEI	38.33	0.23

* the BET surface areas are smaller compared to the calculated theoretical specific surface area due to the high degree of agglomeration during the drying process.

Overall, the thermal decomposition of the introduced bPEI-derivate as a function of temperature was measured with TGA, and ensured the weight mass loss regarding the GOPTS-crosslinked bPEI. The TGA profiles for bare MNPs, MNPs@SiO_2 and MNPs@SiO_2@GOPTS-bPEI are shown in Figure 7a. Bare MNPs display small mass loss (3 wt.%), most likely due to the physisorbed water or residuals from the solvents. A small increase in mass loss was measured for MNPs@SiO_2 (5 wt.%), and can be correlated to the presence of a higher density of –OH groups that release from the silica coating during heating. A very different story can be seen for MNPs@SiO_2@GOPTS-bPEI. The first step in weight loss occurred at around 100 °C, similar as in the case of MNPs@SiO_2, indicating the mass loss due to –OH release. With further heating, the mass loss increased significantly, indicating the GOPTS-bPEI degradation into volatile fragments, and the latter decomposed completely at T = 500 °C. The weight loss regarding the GOPTS-bPEI was estimated to be around 20 wt.%, and is associated with the polymers' thermal degradation into volatile products [44]. With the combining mass loss regarding the GOPTS-bPEI, a rough estimation of the ligand number density per nm^2 can be calculated, assuming that all the excess of reagents was removed by washing and magnetic decantation. Similar ways of ligand density calculations were reported, and the details can be found elsewhere [47]. The value of silanol groups was already reported in literature, and was estimated to be around 5 SiOH nm^{-2} [16]. Ligand density, together with weight loss, accounts to be three molecules of GOPTS-bPEI nm^{-2}. If the assumption of the one primary amino group linkage to the epoxy group is expected (Scheme 1), then three primary amino groups (at pH = 10 full deprotonation of primary amines is expected; refer to the experimental part) could be used for metal-ion chelation. In this context, the density of $-NH_2$ groups per nm^2 can be estimated to be 9 NH_2 nm^{-2}. We can definitely sum up that synthesized nanocomposite exhibiting the surface with amino-rich groups is a promising sorbent for HM via chelation.

The magnetism of bare and differently surface-modified MNPs was explored by room-temperature magnetization curves (Figure 7b). After the coating of the MNPs with a 3 nm thick silica layer, the saturation magnetization decreased from 68 emu·g^{-1} to 33.3 emu·g^{-1} (Figure 7b, plateau part). The dependency of magnetic properties of the amount of the incorporated magnetic phase, shape and size is widely known [14]. Consequently, in nanocomposites constituted from the nonmagnetic and magnetic phases, the amount of incorporated magnetic phase amount plays an important role. In this way, the presence of a nonmagnetic component such as a silica layer reduces the saturated magnetization. With further functionalization of MNPs@SiO_2 with the GOPTS-bPEI, the saturated magnetization of the MNPs@SiO_2@GOPTS-bPEI decreased to 30.7 emu·g^{-1}, due to the additional presence of the nonmagnetic phase originating from the covalently linked amino-rich polymer as an additional layer surrounding the SiO_2-coated MNPs. Among reported values correlated to the saturated magnetization, the reported value herein is one of the largest, just to compare with a few [29,31,48] based on magnetic-bPEI, and surely expresses enough high magnetic force to be applied in real metal-removal application with the synergetic effect of the high density of amino-chelating groups.

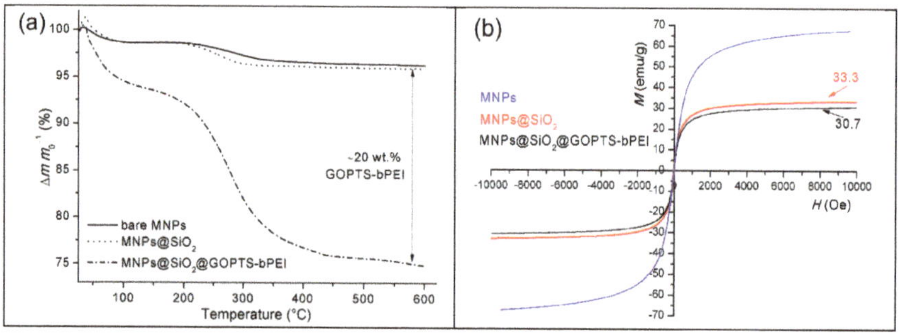

Figure 7. TGA curves for MNPs, MNPs@SiO_2 and MNPs@SiO_2@GOPTS-bPEI (**a**). Room-temperature hysteresis loops for bare MNPs, SiO_2-coated MNPs and MNPs@SiO_2@GOPTS-bPEI (**b**).

3.2. *Effect of the Introduced bPEI onto Cu^{2+} Removal Efficiency*

The adsorption properties of the introduced nanoadsorbent were tested by taking different aspects into account, i.e., (i) The effect of solution pH and (ii) The initial concentration of Cu^{2+}. In both cases, covalently linked amino-rich polymer nanocomposite was compared to MNPs@SiO_2 in order to elucidate the effect of the introduced functional groups. Our preliminary studies (estimating the industry utilization), showed that the the optimal adsorbent (i.e., silica-coated magnetic NPs functionalized with carboxymethylchitosan) dosage was 0.1 mg·mL^{-1} and the removal time 1 hour. Therefore, these two parameters were assumed to be thoroughly optimal and were not examined.

3.2.1. Effect of pH

When taking into consideration the effect of pH, protonation and surface chemistry of the nanoadsorbent, together with the metal ion species distribution, plays an important role (Figure 8). It is known that the Cu^{2+} starts to form copper hydroxide precipitates at pH > of 6 [49]; therefore, the influence of the pH on the adsorption capacity was studied at pH = 6. Electrokinetic measurements (Figure 5a) showed that the IEP is located at around pH 8.5. Regarding this and the determined pK values for primary groups of bPEI (Figure 5b), it can be supposed that, at pH = 6, most of the primary amines are in deprotonated form. Therefore, they are highly accessible, due to the electron-donor pair with favorable Cu^{2+} chelation [50]. Therefore, larger Cu^{2+} adsorption capacity can be seen at pH = 6 when taking into account that the copper uptake mechanism most likely occurs through deprotonated primary groups [51], indicating the higher sorption of the Cu^{2+}. At lower pH (pH < 6), the adsorption capacity is almost constant (~60 mg·g^{-1}). This also correlates well with the pH-dependent titration and ZP behavior (Figure 4a,b), where most of the amino functional groups are in protonated state. Therefore, due to the presence of $-NH_3^+$ groups, the competition of H^+ with the Cu^{2+} for binding sites can also be present. As already reported, the latter is more pronounced at lower pH values [17].

Beside this, the electrostatic repulsions between both positively charged MNPs@SiO_2@GOPTS-bPEI and Cu^{2+} can, in part, hinder the Cu^{2+} adsorption. Different behavior can be observed for MNPs@SiO_2. At pH = 3, the adsorption capacity of MNPs@SiO_2 is approximately 5 mg·g^{-1}. Adsorption efficiency increased with rising pH, most likely due to the electrostatic interactions between the negatively charged silanol groups and Cu^{2+}. Nevertheless, the adsorption capacity of the GOPTS-bPEI-based adsorbent was 6-times higher at pH = 6 with respect to MNPs@SiO_2 (Figure 8), indicating the advantage of covalently introduced derived-bPEI polymer.

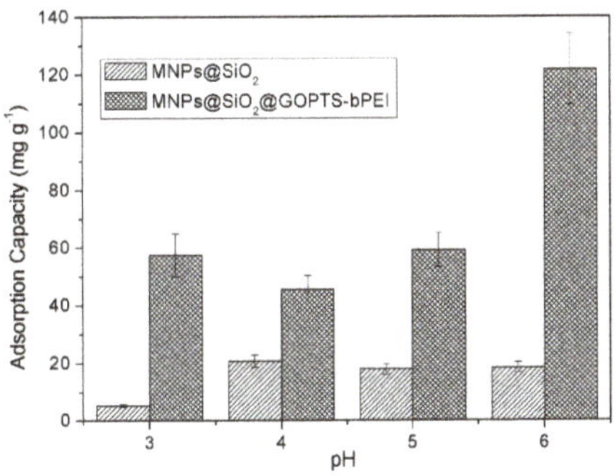

Figure 8. Influence of solution pH on adsorption capacity of Cu^{2+} for MNPs@SiO_2 and GOPTS-bPEI modified MNPs@SiO_2.

3.2.2. Effect of Initial Cu^{2+} Concentration

The influence was evaluated of initial Cu^{2+} concentration on adsorption efficiency (Figure 9). Adsorption capacity increased with upturning of the Cu^{2+} solution to the plateau part, where the Cu^{2+} in the solution is in balance to the Cu^{2+} onto the adsorbent, occupying all the amino groups. This may be the result of the difference in the concentration gradient between Cu^{2+} in the initial solution and its absence on the nanoadsorbent, acting as a driving force until all the active sorption places are overtaken. The obtained results were well fitted with the Langmuir isotherm (inset in Figure 9), suggesting the monolayer coverage of Cu^{2+} onto the nanoadsorbent during the adsorption process. The maximum adsorption capacity was shown to be 143 mg·g^{-1} and is one of the largest in comparison to the few previously studied PEI-based adsorbents for Cu^{2+} removal applications [44,51–55]. The latter can be attributed to the high density, and to the homogeneous distribution of amino groups from bPEI, as well as to the high specific surface area and porous structure. The role of surface modification was additionally compared to MNPs@SiO_2 adsorption tendencies (Figure 9) and clearly elucidates the significant sorption capabilities with covalently introduced GOPTS-bPEI for HM removal.

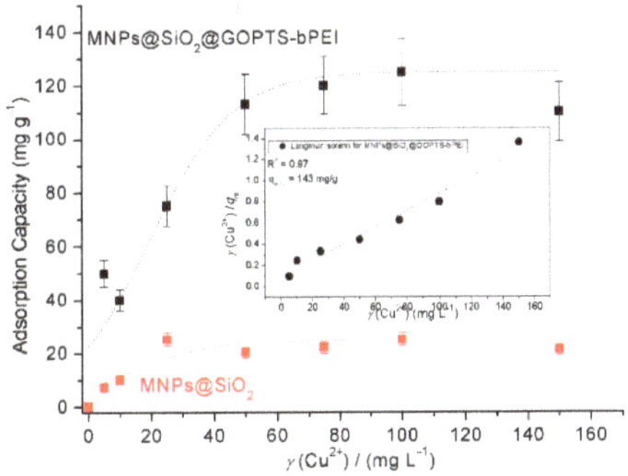

Figure 9. Effect of the initial Cu^{2+} concentration on the adsorption capacity for unmodified and GOPTS-bPEI modified MNPs@SiO_2.

3.2.3. Reusability Cycles for Regeneration Studies

Regeneration is an important indicator of nanoadsorbent economical attractiveness that contributes additionally to the waste minimization concept, where the same adsorbent can be used several times. It is determined mainly via adsorption–desorption experiments [24], and was calculated with Equation 2. In order to achieve effective desorption of HMs, and to apply adsorbents further to the next adsorption process, eluents with high desorption ability are exceptionally desirable. In our case, Na_2EDTA was considered as the most appropriate eluent for regeneration studies, whilst it has already shown high desorption capability of the Na_2EDTA usage as an eluent in the case of Cu^{2+} [38]. Figure 10 shows that the removal efficiency of the amino-rich nanoadsorbent decreased only to 95% after the second reusability cycle. MNPs@SiO_2@GOPTS-bPEI still showed high adsorption tendencies for Cu^{2+} (88% of removal efficiency, Figure 10), even after the fourth adsorption–desorption cycle. This indicates possible multiple regeneration cycles of the nanoadsorbent, together with superior removal efficiency.

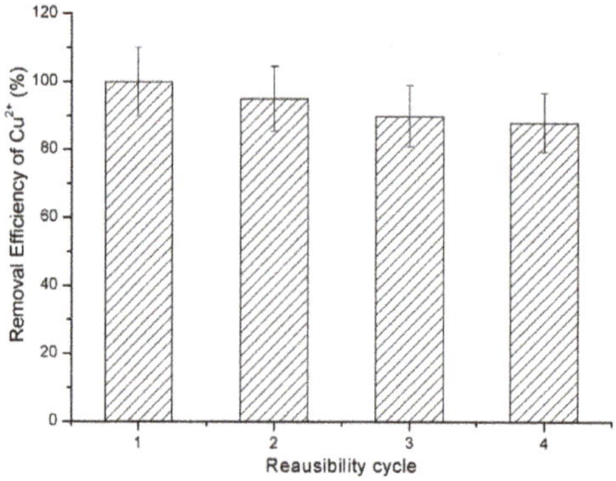

Figure 10. Reusability cycles of MNPs@SiO_2@GOPTS-bPEI, expressed as adsorption–desorption studies, for copper removal.

3.3. Proposed Possible Adsorption Mechanism for Cu

The adsorption mechanisms of Cu onto nanoadsorbent were explained using different surface analytical techniques. XPS surface technique revealed the presence of Cu after adsorption at the MNPs@SiO_2@GOPTS-bPEI surface (Figure 11), underscoring its uptake after sorption. The surface composition before and after Cu uptake was normalized relative to oxygen (Figure 11a). It can be observed that the surface composition did not change significantly after Cu adsorption, only the presence of two additional elements was detected (i.e., Cu and Cl). Cu originates as a consequence of adsorption, while Cl originates from the copper model solution, and is most likely attached to MNPs@SiO_2@GOPTS-bPEI via electrostatic attraction, despite magnetic separation and washing. Similar surface composition before and after Cu uptake shows clearly nanoadsorbent stability, and no desorption of the introduced derived amino-polymer. The high-resolution core spectrum for Cu 2p is shown in Figure 11b. The copper peak Cu $2p_{3/2}$ is located at the binding energy 933 eV, and can be correlated to Cu^0 along with Cu^+, as both are present at the similar bonding energy, and it is difficult to distinguish between them. However, the presence of Cu^{2+} after uptake can be excluded, as the typical peak for Cu^{2+} should be present at 933.5–934 eV. Moreover, typical satellite peaks in the area 940–945 eV should be observed [42]. None of these were seen in our case. The reduced electronic state of Cu can be attributed to the filling of the empty Cu atomic orbital with free electron pairs from the deprotonated amino groups [51]. This is in agreement with the Langmuir sorption model fits (inset in Figure 9), indicating monolayer adsorption, and suggesting that the adsorption mechanism corresponds to the chemisorption, together with the prediction that the adsorption occurred mainly due to the complexation of the Cu ions with the bPEI amine groups. Additionally, the authors in Reference [56] proposed calculations for different models that predict the manner of the metal ion bonding to the amine active adsorption sites. Taking their calculation with our parameters into consideration revealed the "pendant model", which indicates that the chelation of copper with only one amino group is more suitable to describe the mechanism of Cu adsorption onto MNPs@SiO_2@GOPTS-bPEI in the present study.

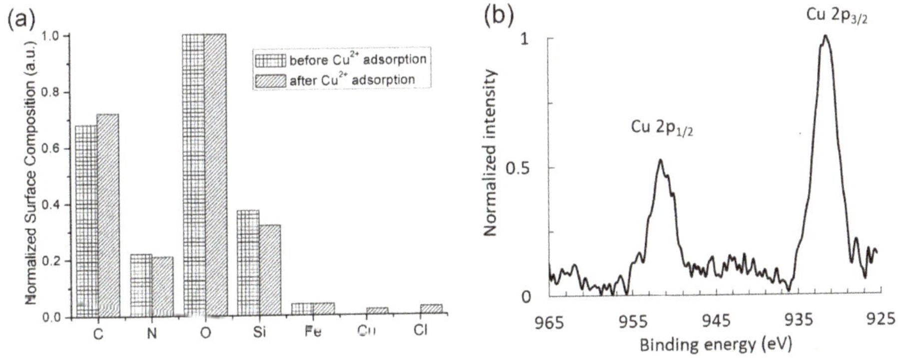

Figure 11. Normalized surface composition before and after Cu adsorption of MNPs@SiO_2@GOPTS-bPEI (normalized to oxygen) is shown in (**a**) while the high-resolution core spectra for Cu 2p is shown in (**b**).

The MNPs@SiO_2@GOPTS-bPEI were used as a Cu adsorbent at optimal conditions, and Cu uptake was characterized in terms of ATR-FTIR (Figure 12a). It was clearly observed that the major changes after Cu adsorption varied at the wavenumbers, typical for –NH vibrations. This was confirmed additionally with subtracting the FTIR spectra before and after Cu adsorption (inset in Figure 12a). With respect to that fact, it can be assumed that these functional groups are involved during Cu adsorption (complexation mechanism), and cause changes in vibrations of these functional groups. Similar observations were reported in Reference [17], and also support the observation from the XPS analysis above.

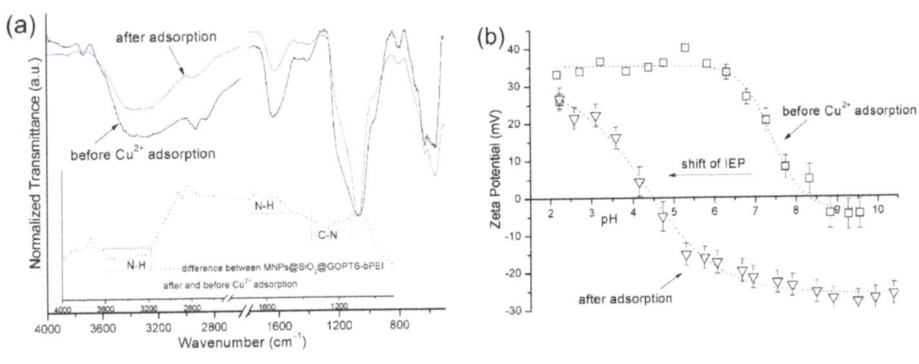

Figure 12. Attenuated Total Reflection (ATR) FTIR spectra before and after Cu adsorption with corresponding difference between them (**a**), and Zeta Potential behavior as a pH function for MNPs@SiO_2@GOPTS-bPEI before and after Cu adsorption (**b**).

The electrokinetic measurements of the ZP were directly correlated to the surface potential behavior after Cu uptake by the amino-nanoadsorbent (Figure 12b). ZP measurements in a wide pH range showed the shift of the IEP after Cu adsorption to lower pH values (pH 4.5) in comparison to the pure adsorbent (pH 8.5). It can be assumed that the Cu is sharing free electron pairs with the amino groups from MNPs@SiO_2@GOPTS-bPEI, and the latter are not accessible anymore for the protonation to express a highly positive charge. Moreover, when Cu is adsorbed onto MNPs@SiO_2@GOPTS-bPEI, its more acidic character may be connected to the presence of Cu acting as an acid [57].

4. Conclusions

This paper represents for the first time the systematic concept of crosslinking epoxy-containing organosilane with branched PEI through a ring-opening reaction, and its further silanization onto silica-coated MNPs for copper removal applications through possible chelation mechanisms via deprotonated amino groups. The synthesized nanoadsorbent was composed of 13 nm-sized magnetic cores, uniformly covered with an approximately 3-nm-thick amorphous silica layer, and additionally silanized with GOPTS-derived bPEI. FTIR analysis revealed the presence of typical functional groups after silanization, as well as the XPS and TGA showed high content of N, suggesting an -NH_2 rich adsorbent surface. Moreover, correlations between electrokinetic measurements and pH-potentiometric titrations were in high agreement with the proposed bonding mechanism. Additionally, the large magnetic response and precisely managed thickness of the surface layers showed quick removal after the finished adsorption process, which is highly appreciated for a successful adsorption process. In order to evaluate the efficiency of the introduced adsorbent, the effect of initial pH and the initial Cu^{2+} solution concentration were studied, and compared with respect to MNPs@SiO_2. Results revealed the importance of surface modification, pH-dependency of the copper adsorption, and the monolayer coverage with large adsorption capacity. Moreover, after reusing the adsorbent, the latter shows high tendency towards Cu removal, even after several cycles. The proposed adsorption mechanism revealed major involvement of the NH/NH_2 functional groups by the adsorption of copper by reducing its electronic state by donating free electron pairs.

To conclude, the developed amino-rich nanoadsorbent represents an efficient, environmentally friendly, stable and straightforwardly prepared adsorbent for metal removal, especially in more complicated environmental conditions, such as sludge.

Author Contributions: Conceptualization, O.P. and L.F.Z.; methodology, O.P., M.F., L.F.Z., S.G., U.A., I.B.; investigation, O.P.; writing—original draft preparation, O.P., M.F., L.F.Z.; writing—review and editing, O.P., M.F., S.G., U.A., I.B., L.F.Z.; supervision, L.F.Z.

Funding: This research received no external funding.

Acknowledgments: The financial support of the Ministry of Education, Science and Sport of the Republic of Slovenia under Contract C3330-17-529004, and from the Slovenian Research Agency (Research Core Funding Nos. P2-0118, P2–0089 and P2-0032) is gratefully acknowledged. The authors also acknowledge the use of equipment in the Center of Excellence on Nanoscience and Nanotechnology—Nanocenter. We also acknowledge Dr. Matej Bračič, Anja Petek, Kristina Olovec and Dr. Alenka Vesel for their support regarding the analyses and laboratory work.

Conflicts of Interest: The authors declare no conflicts of interest.

References

1. European Commission. Protection of the Environment, and in particular of the soil, when sewage sludge is used in agriculture. *Off. J. Eur. Communities* **1986**, *4*, 6–12.
2. Babel, S.; del Mundo Dacera, D. Heavy metal removal from contaminated sludge for land application: A review. *Waste Manag.* **2006**, *26*, 988–1004. [CrossRef] [PubMed]
3. Ahmed, H.; Fawy, H.; Abdel-Hady, E. Study of sewage sludge use in agriculture and its effect on plant and soil. *Agric. Biol. J. N. Am.* **2010**, *1*, 1044–1049. [CrossRef]
4. Bratina, B.; Šorgo, A.; Kramberger, J.; Ajdnik, U.; Zemljič, L.F.; Ekart, J.; Šafarič, R. From municipal/industrial wastewater sludge and FOG to fertilizer: A proposal for economic sustainable sludge management. *J. Environ. Manag.* **2016**, *183*, 1009–1025. [CrossRef] [PubMed]

5. Pradel, M.; Aissani, L.; Villot, J.; Baudez, J.C.; Laforest, V. From waste to added value product: Towards a paradigm shift in life cycle assessment applied to wastewater sludge—A review. *J. Clean. Prod.* **2016**, *131*, 60–75. [CrossRef]
6. Zhou, C.; Ge, S.; Yu, H.; Zhang, T.; Cheng, H.; Sun, Q.; Xiao, R. Environmental risk assessment of pyrometallurgical residues derived from electroplating and pickling sludges. *J. Clean. Prod.* **2018**, *177*, 699–707. [CrossRef]
7. Gaetke, L.M.; Chow, C.K. Copper toxicity, oxidative stress, and antioxidant nutrients. *Toxicology* **2003**, *189*, 147–163. [CrossRef]
8. Hua, M.; Zhang, S.; Pan, B.; Zhang, W.; Lv, L.; Zhang, Q. Heavy metal removal from water/wastewater by nanosized metal oxides: A review. *J. Hazard. Mater.* **2012**, *211–212*, 317–331. [CrossRef]
9. Ali, I.; Gupta, V.K. Advances in water treatment by adsorption technology. *Nat. Protoc.* **2007**, *1*, 2661–2667. [CrossRef]
10. Mudhoo, A.; Garg, V.K.; Wang, S. Removal of heavy metals by biosorption. *Environ. Chem. Lett.* **2012**, *10*, 109–117. [CrossRef]
11. Das, R.; Vecitis, C.D.; Schulze, A.; Cao, B.; Ismail, A.F.; Lu, X.; Chen, J.; Ramakrishna, S. Recent advances in nanomaterials for water protection and monitoring. *Chem. Soc. Rev.* **2017**, *46*, 6946–7020. [CrossRef] [PubMed]
12. Kong, A.; Ji, Y.; Ma, H.; Song, Y.; He, B.; Li, J. A novel route for the removal of Cu(II) and Ni(II) ions via homogeneous adsorption by chitosan solution. *J. Clean. Prod.* **2018**, *192*, 801–808. [CrossRef]
13. Jung, J.H.; Lee, J.H.; Shinkai, S. Functionalized magnetic nanoparticles as chemosensors and adsorbents for toxic metal ions in environmental and biological fields. *Chem. Soc. Rev.* **2011**, *40*, 4464. [CrossRef] [PubMed]
14. Cornell, R.M.; Schwertmann, U. *The Iron Oxides: Structure, Reactions, Occurrences and Uses*, 2nd ed.; WILEY-VCH: Weinheim, Germany, 2003.
15. Kralj, S.; Makovec, D.; Čampelj, S.; Drofenik, M. Producing ultra-thin silica coatings on iron-oxide nanoparticles to improve their surface reactivity. *J. Magn. Magn. Mater.* **2010**, *322*, 1847–1853. [CrossRef]
16. Iler, R.K. *The Chemistry of Silica: Solubility, Polymerization, Colloid and Surface Properties, and Biochemistry*; John Wiley&Sons: New York, NY, USA, 1979; ISBN 047102404X.
17. Pang, Y.; Zeng, G.; Tang, L.; Zhang, Y.; Liu, Y.; Lei, X.; Li, Z.; Zhang, J.; Xie, G. PEI-grafted magnetic porous powder for highly effective adsorption of heavy metal ions. *Desalination* **2011**, *281*, 278–284. [CrossRef]
18. Sun, W. Functionalization of surfaces with branched polymers. *RSC Adv.* **2016**, *6*, 42089–42108. [CrossRef]
19. Virgen-Ortíz, J.J.; dos Santos, J.C.; Berenguer-Murcia, Á; Barbosa, O.; Rodrigues, R.C.; Fernandez-Lafuente, R. Polyethylenimine: A very useful ionic polymer in the design of immobilized enzyme biocatalysts. *J. Mater. Chem. B* **2017**, *5*, 7461–7490.
20. Kim, C.; Lee, S.S.; Lafferty, B.J.; Giammar, D.E.; Fortner, J.D. Engineered superparamagnetic nanomaterials for arsenic (v) and chromium (vi) sorption and separation: Quantifying the role of organic surface coatings. *Environ. Sci. Nano* **2018**, *5*, 556–563. [CrossRef]
21. Goon, I.Y.; Zhang, C.; Lim, M.; Gooding, J.J.; Amal, R. Controlled fabrication of polyethylenimine-functionalized magnetic nanoparticles for the sequestration and quantification of free Cu^{2+}. *Langmuir* **2010**, *26*, 12247–12252. [CrossRef]
22. Wang, Y.; Wu, D.; Wei, Q.; Wei, D.; Yan, T.; Yan, L.; Hu, L.; Du, B. Rapid removal of Pb (II) from aqueous solution using branched polyethylenimine enhanced magnetic carboxymethyl chitosan optimized with response surface methodology. *Sci. Rep.* **2017**, *7*, 1–11. [CrossRef]
23. Liu, M.; Zhang, B.; Wang, H.; Zhao, F.; Chen, Y.; Sun, Q. Facile crosslinking synthesis of hyperbranch-substrate nanonetwork magnetite nanocomposite for the fast and highly efficient removal of lead ions and anionic dyes from aqueous solutions. *RSC Adv.* **2016**, *6*, 67057–67071. [CrossRef]
24. Wu, Y.; Chen, L.; Long, X.; Zhang, X.; Pan, B.; Qian, J. Multi-functional magnetic water purifier for disinfection and removal of dyes and metal ions with superior reusability. *J. Hazard. Mater.* **2018**, *347*, 160–167. [CrossRef] [PubMed]
25. Wong, S.; Tumari, H.H.; Ngadi, N.; Mohamed, N.B.; Hassan, O.; Mat, R.; Aishah, N.; Amin, S. Adsorption of anionic dyes on spent tea leaves modified with polyethyleneimine (PEI-STL). *J. Clean. Prod.* **2019**, *206*, 394–406. [CrossRef]
26. Hermanson, T.G. *Bioconjugate Techniques*, 2nd ed.; Elsevier: Amsterdam, The Netherlands, 2008; ISBN 9780123705013.

27. Larraza, I.; López-Gónzalez, M.; Corrales, T.; Marcelo, G. Hybrid materials: Magnetite-Polyethylenimine-Montmorillonite, as magnetic adsorbents for Cr(VI) water treatment. *J. Colloid Interface Sci.* **2012**, *385*, 24–33. [CrossRef] [PubMed]
28. Jiang, H.; Sun, M.; Xu, J.; Lu, A.; Shi, Y. Magnetic Fe3O4 Nanoparticles Modified With Polyethyleneimine for the Removal of Pb(II). *Clean Soil Air Water* **2016**, *44*, 1–8. [CrossRef]
29. Sui, N.; Wang, L.; Wu, X.; Li, X.; Sui, J.; Xiao, H.; Liu, M.; Wan, J.; Yu, W.W. Polyethylenimine modified magnetic graphene oxide nanocomposites for Cu^{2+} removal. *RSC Adv.* **2015**, *5*, 746–752. [CrossRef]
30. Zhong, L.; Zhang, Q.; Sun, M.; Zhang, Y.; Jiang, H.; Lian, H. Fabrication and characterization of polyethyleneimine immobilized on chloropropyl- and silica-coated magnetic nanoparticles for Pb^{2+} removal from aqueous solution. *Desalin. Water Treat.* **2016**, *57*, 13701–13710. [CrossRef]
31. Sun, X.; Yang, L.; Xing, H.; Zhao, J.; Li, X.; Huang, Y.; Liu, H. Synthesis of polyethylenimine-functionalized poly(glycidyl methacrylate) magnetic microspheres and their excellent Cr(VI) ion removal properties. *Chem. Eng. J.* **2013**, *234*, 338–345. [CrossRef]
32. Sun, X.; Yang, L.; Li, Q.; Liu, Z.; Dong, T.; Liu, H. Polyethylenimine-functionalized poly(vinyl alcohol) magnetic microspheres as a novel adsorbent for rapid removal of Cr(VI) from aqueous solution. *Chem. Eng. J.* **2015**, *262*, 101–108. [CrossRef]
33. Chen, B.; Zhao, X.; Liu, Y.; Xu, B.; Pan, X. Highly stable and covalently functionalized magnetic nanoparticles by polyethyleneimine for Cr(VI) adsorption in aqueous solution. *RSC Adv.* **2015**, *5*, 1398–1405. [CrossRef]
34. Li, W.; Liu, Q.; Chen, R.; Yu, J.; Zhang, H.; Liu, J.; Li, R.; Zhang, M.; Liu, P.; Wang, J. Efficient removal of U(vi) from simulated seawater with hyperbranched polyethylenimine (HPEI) covalently modified SiO_2 coated magnetic microspheres. *Inorg. Chem. Front.* **2018**, *5*, 1321–1328. [CrossRef]
35. Campelj, S.; Makovec, D.; Drofenik, M. Preparation and properties of water-based magnetic fluids. *J. Phys. Condens. Matter* **2008**, *20*, 204101. [CrossRef]
36. Čakara, D.; Fras, L.; Bračič, M.; Kleinschek, K.S. Protonation behavior of cotton fabric with irreversibly adsorbed chitosan: A potentiometric titration study. *Carbohydr. Polym.* **2009**, *78*, 36–40. [CrossRef]
37. Rinaudo, M. Chitin and chitosan: Properties and applications. *Prog. Polym. Sci.* **2006**, *31*, 603–632. [CrossRef]
38. Zhou, Y.-T.; Nie, H.-L.; Branford-White, C.; He, Z.-Y.; Zhu, L.-M. Removal of Cu2+ from aqueous solution by chitosan-coated magnetic nanoparticles modified with α-ketoglutaric acid. *J. Colloid Interface Sci.* **2009**, *330*, 29–37. [CrossRef] [PubMed]
39. Van Quy, D.; Hieu, N.M.; Tra, P.T.; Nam, N.H.; Hai, N.H.; Thai Son, N.; Nghia, P.T.; Van Anh, N.T.; Hong, T.T.; Luong, N.H. Synthesis of silica-coated magnetic nanoparticles and application in the detection of pathogenic viruses. *J. Nanomater.* **2013**, *2013*, 603940.
40. Motevalizadeh, S.F.; Khoobi, M.; Sadighi, A.; Khalilvand-Sedagheh, M.; Pazhouhandeh, M.; Ramazani, A.; Faramarzi, M.A.; Shafiee, A. Lipase immobilization onto polyethylenimine coated magnetic nanoparticles assisted by divalent metal chelated ions. *J. Mol. Catal. B Enzym.* **2015**, *120*, 75–83. [CrossRef]
41. Larkin, P.J. *IR and Raman Spectroscopy—Principles and Spectral Interpretation*; Elsevier: Amsterdam, The Netherlands, 2011; ISBN 9780123869845.
42. Finšgar, M. 2-Mercaptobenzimidazole as a copper corrosion inhibitor: Part II. Surface analysis using X-ray photoelectron spectroscopy. *Corros. Sci.* **2013**, *72*, 90–98. [CrossRef]
43. Finšgar, M.; Fassbender, S.; Hirth, S.; Milošev, I. Electrochemical and XPS study of polyethyleneimines of different molecular sizes as corrosion inhibitors for AISI 430 stainless steel in near-neutral chloride media. *Mater. Chem. Phys.* **2009**, *116*, 198–206. [CrossRef]
44. Huang, Q.; Liu, M.; Zhao, J.; Chen, J.; Zeng, G.; Huang, H.; Tian, J.; Wen, Y.; Zhang, X.; Wei, Y. Facile preparation of polyethylenimine-tannins coated SiO_2 hybrid materials for Cu_{2+} removal. *Appl. Surf. Sci.* **2018**, *427*, 535–544. [CrossRef]
45. Sing, K.S.W. Adsorption methods for the characterization of porous materials. *Adv. Colloid Interface Sci.* **1998**, *76–77*, 3–11. [CrossRef]
46. Lee, M.Y.; Lee, J.H.; Chung, J.W.; Kwak, S.Y. Hydrophilic and positively charged polyethylenimine-functionalized mesoporous magnetic clusters for highly efficient removal of Pb(II) and Cr(VI) from wastewater. *J. Environ. Manag.* **2018**, *206*, 740–748. [CrossRef]
47. Tong, L.; Lu, E.; Pichaandi, J.; Cao, P.; Nitz, M.; Winnik, M.A. Quantification of Surface Ligands on $NaYF_4$ Nanoparticles by Three Independent Analytical Techniques. *Chem. Mater.* **2015**, *27*, 4899–4910. [CrossRef]

48. Xiao, C.; Liu, X.; Mao, S.; Zhang, L.; Lu, J. Sub-micron-sized polyethylenimine-modified polystyrene/Fe_3O_4/chitosan magnetic composites for the efficient and recyclable adsorption of Cu(II) ions. *Appl. Surf. Sci.* **2017**, *394*, 378–385. [CrossRef]
49. Baes, C.F.; Mesmer, R.E. *The Hydrolysis of Cations*; John Wiley & Sons, Inc.: Toronto, ON, Canada, 1976; ISBN 0471039853.
50. Perrine, T.D.; Landis, W.R. Analysis of polyethylenimine by spectrophotometry of its copper chelate. *J. Polym. Sci. A1* **1967**, *5*, 1993–2003. [CrossRef] [PubMed]
51. Wang, J.; Li, Z. Enhanced selective removal of Cu(II) from aqueous solution by novel polyethylenimine-functionalized ion imprinted hydrogel: Behaviors and mechanisms. *J. Hazard. Mater.* **2015**, *300*, 18–28. [CrossRef] [PubMed]
52. Tang, J.; Xiang, L.; Fan, J.; Cui, Z.; Wu, H. Synthesis of polyethylenimine-functionalized magnetic materials and a critical evaluation of the removal of copper in sediments. *Environ. Earth Sci.* **2018**, *77*, 441. [CrossRef]
53. Zhang, N.; Zang, G.L.; Shi, C.; Yu, H.Q.; Sheng, G.P. A novel adsorbent TEMPO-mediated oxidized cellulose nanofibrils modified with PEI: Preparation, characterization, and application for Cu(II) removal. *J. Hazard. Mater.* **2016**, *316*, 11–18. [CrossRef] [PubMed]
54. Huang, Z.; Huang, Z.; Feng, L.; Luo, X.; Wu, P.; Cui, L.; Mao, X. Modified cellulose by polyethyleneimine and ethylenediamine with induced Cu(II) and Pb(II) adsorption potentialities. *Carbohydr. Polym.* **2018**, *202*, 470–478. [CrossRef]
55. Huang, J.; Xu, Y.; Zhang, X.; Lei, Z.; Chen, C.; Deng, Y.; Wang, C. Polyethylenimine and dithiocarbamate decorated melamine sponges for fast copper (II) ions removal from aqueous solution. *Appl. Surf. Sci.* **2018**, *445*, 471–477. [CrossRef]
56. Yu, K.; Ho, J.; McCandlish, E.; Buckley, B.; Patel, R.; Li, Z.; Shapley, N.C. Copper ion adsorption by chitosan nanoparticles and alginate microparticles for water purification applications. *Colloids Surf. A Physicochem. Eng. Asp.* **2013**, *425*, 31–41. [CrossRef]
57. Taboada, E.; Cabrera, G.; Cardenas, G. Retention Capacity of Chitosan for Copper and Mercury Ions. *J. Chil. Chem. Soc.* **2003**, *48*, 7–12. [CrossRef]

© 2019 by the authors. Licensee MDPI, Basel, Switzerland. This article is an open access article distributed under the terms and conditions of the Creative Commons Attribution (CC BY) license (http://creativecommons.org/licenses/by/4.0/).

nanomaterials

MDPI

Article

Efficient Removal of Lead, Copper and Cadmium Ions from Water by a Porous Calcium Alginate/Graphene Oxide Composite Aerogel

Linhai Pan [1], Zhuqing Wang [1,2,*], Qi Yang [1] and Rongyi Huang [1]

[1] AnHui Provice Key Laboratory of Optoelectronic and Magnetism Functional Materials, Anqing Normal University, Anqing 246011, China; panlinhai@nimte.ac.cn (L.P.); wangzhq@aqnu.edu.cn (Q.Y.); huangry@aqnu.edu.cn (R.H.)

[2] Faculty of Production Engineering, University of Bremen, Am Fallturm 1, D-28359 Bremen, Germany

* Correspondence: wzhuqing@uni-bremen.de; Tel.: +49-421-218-64581

Received: 1 November 2018; Accepted: 17 November 2018; Published: 20 November 2018

Abstract: In this study, we fabricated a porous calcium alginate/graphene oxide composite aerogel by using polystyrene colloidal particles as sacrificial template and graphene oxide as a reinforcing filler. Owing to the excellent metal chelation ability of calcium alginate and controlled nanosized pore structure, the as-prepared calcium alginate/graphene oxide composite aerogel (mp-CA/GO) can reach the adsorption equilibrium in 40 min, and the maximum adsorption capacity for Pb^{2+}, Cu^{2+} and Cd^{2+} is 368.2, 98.1 and 183.6 mg/g, respectively. This is higher than most of the reported heavy metal ion sorbents. Moreover, the mp-CA/GO can be regenerated through simple acid-washing and be used repeatedly with little loss in performance. The adsorption mechanism analysis indicates that the mp-CA/GO adsorb the heavy metal ions mainly through the ion exchange and chemical coordination effects.

Keywords: Adsorbent; heavy metal ions; calcium alginate; graphene oxide; polystyrene colloidal particles template

1. Introduction

Heavy metal pollution is currently a serious environmental problem. Heavy metal ions are not biodegradable and tend to accumulate in living organisms, thereby causing many serious health and environment problems [1,2]. Therefore, it is of great significance to remove heavy metal ions in water. In recent years, many methods such as ion-exchange, chemical oxidation and reduction, membrane filtration, chemical precipitation, and adsorption have been employed for removing heavy metal ions from water media [3–7]. Among these treatment methods, adsorption attracts more attention due to its simple operation, low price, recyclability of the adsorbent, and high efficiency in treating low-concentration wastewater [8].

The exploration of efficient adsorbent materials starts by providing additional binding sites for adsorbed ions, improving the diffusion coefficient during the adsorption process, and enhancing the overall mechanical properties of adsorbent materials. Adsorbents with a large specific surface area, additional active groups, such as carboxyl and hydroxyl groups, and rational diffusion path for metal ion during adsorption are required. Deze et al. demonstrated the effect of porosity in heavy metal ions sorption [9].

In our previous work, we synthesized pure calcium alginate and chitosan-calcium alginate hybrid aerogels for removing heavy metal ions in water [10,11]. The results showed that calcium alginate could efficient removal of Pb^{2+} and Cu^{2+} from wastewater due to its rich carboxyl (−COOH) and hydroxyl (−OH) groups. However, dry alginate bead that has been used for heavy metal ion sorption is a rigid material without controlled porous structure, and given the low diffusion coefficient of the material, the sorption kinetics is relatively slow. Furthermore, pure alginate also exhibits several unsatisfactory structural properties, such as weak mechanical strength, structural instability, and fragile collapse [12], thus limiting their applications in the actual setting.

Graphene oxide (GO) is a new carbon material with excellent properties, high specific surface area, and rich surface functional groups. GO also exhibits a considerable potential in reinforcing fillers given its outstanding mechanical properties, excellent binding capacity, and superior flexibility [13]. The unsatisfactory properties of alginate structure and collapse of a porous material structure can be easily solved by adding GO sheets as ideal reinforcing fillers for composites. Jiao et al. presented that GO, as a reinforcing filler, exhibits excellent mechanical strength and elasticity in adsorbing heavy metal ions [13]. Recently, Yang et al. successfully prepared double network hydrogel beads by directly mixing GO and sodium alginate solutions. The as-prepared hydrogel beads showed good affinity to cationic metals and the theoretical maximum adsorption capacity for Mn^{2+} reached 56.49 mg/g [14]. However, preparing adsorbent materials with controlled pore structure and excellent mechanical properties has not been reported.

In this study, a novel macro-porous (mp) calcium alginate/graphene oxide composite aerogel (mp-CA/GO) with controlled pore structure was prepared by introducing macropores within the composite aerogel using polystyrene (PS) colloidal particles as sacrificial template and GO as reinforcing fillers. The as-prepared mp-CA/GO was characterized and used for heavy metal ions (Pb^{2+}, Cu^{2+} and Cd^{2+}) sorption. In addition, adsorption capacity, kinetics and thermodynamics properties, adsorption mechanism, and reutilization were also explored.

2. Materials and Methods

2.1. Materials

Sodium alginate, ethanol, styrene, potassium persulfate ($K_2S_2O_8$), nitric acid (HNO_3), GO suspension (in water), sodium chloride (NaCl), and other metal salts were bought from Shanghai Aladdin Biochemical Technology Co. Ltd. (Shanghai, China). Certain reagents such as styrene, potassium persulfate, and sodium chloride, were used with further purification. High concentration GO suspension was diluted with deionized (DI) water, and DI water was used throughout this study.

2.2. Instruments

Scanning electron microscopy (SEM) images were obtained on a Sirion200 microscope (FEI Company, Eindhoven, the Netherlands) at an accelerating voltage of 10.0 kV. The infrared spectra were obtained from a Nicolet 6700 Fourier transform infrared (FT-IR) spectrometer (Thermo Fisher Scientific, Waltham, MA, USA). X-ray diffraction (XRD) pattern was collected on a Bruker axs D8 advanced diffractometer (Bruker Corporation, Frankfurt, Germany) using Cu Kα radiation. X-ray photoelectron spectroscopy (XPS) spectra were collected on a Shimadzu Axis-Ultra multifunctional X-ray photoelectron spectrometer (Shimadzu Corporation, Tokyo, Japan) using an Al K X-ray source. Metal concentrations were confirmed using a Perkin-Elmer Optima 2100 (Perkin-Elmer Company, Waltham, MA, USA) inductively coupled plasma-optical emission spectrometry (ICP-OES).

2.3. Preparation of PS Colloidal Particles

Monodisperse PS colloidal particles were synthesized by soapless emulsion polymerization [15]. Styrene is purified by vacuum distillation before using 150 mL of DI water, 0.175 g sodium chloride and 25 g styrene were added into a 250 mL three-port bottle, and stirred in the water bath at 70 °C. Then, 0.2 g of $K_2S_2O_8$ was added to the bottle after ventilating nitrogen for nearly 20 min. With nitrogen protection, the polymerization reaction was conducted at 70 °C for 20 h. After the reaction, the particles were subjected to repeated centrifugal sedimentation at 9000 rpm and ultrasonic dispersion in water and ethanol to remove the styrene monomer and sodium chloride. Finally, the PS particles were mixed into a solution with 10% mass fraction.

2.4. Preparation of mp-CA/GO

For the fabrication of the mp-CA/GO, the GO (0.5 mg mL^{-1}, 2 mL) and sodium alginate (1% w/v, 20 mL) solutions were mixed homogeneously. Then, PS colloidal particles solution (0.5 mL, 10 wt.%) was added to the sodium alginate and GO mixed solutions under magnetic stirring to form a homogeneous mixture. After that, this mixed solution was added dropwise to 100 mL of 0.2 M Ca^{2+} solution. A hydrogel sphere formed immediately when the mixed solution contacted with the Ca^{2+} solution. The hydrogel spheres were collected and washed with DI water, and placed into a low-temperature freezer. Then, the frozen hydrogel spheres were freeze-dried for 24 h under vacuum. The final mp-CA/GO composite aerogel were obtained by removing the PS colloidal particles with toluene and tetrahydrofuran exposure through ultrasonication [16].

2.5. Adsorption and Desorption Tests

In an adsorption test, ~ 50 mg of mp-CA/GO was added to 50 mL of 1.5 mM Pb^{2+}, Cu^{2+}, and Cd^{2+} solutions, respectively. The mixture was filtered after stirring for 40 min. The unextracted Pb^{2+}, Cu^{2+}, and Cd^{2+} in the filtrate were determined by ICP-OES.

In the desorption test, the Pb^{2+}-, Cu^{2+}-, or Cd^{2+}-loaded mp-CA/GO was first immersed in 50 mL of 0.07 M HNO_3 solution for 20 min. Then, the mp-CA/GO was separated and washed with DI water. The desorbed Pb^{2+}, Cu^{2+}, and Cd^{2+} in the eluent were also determined by ICP-OES. Each adsorption or desorption test was performed three times in parallel. The metal ion concentration was also determined three times in parallel and then averaged.

The adsorption capacity (Q, mg/g) and Adsorption (%) were calculated as follows:

$$Q = \frac{\left(C_i - C_f\right) \cdot V \cdot M}{m} \tag{1}$$

$$\text{Adsorption} = \frac{\left(C_i - C_f\right)}{C_i} \times 100\% \tag{2}$$

where V is the volume of the solution (L); m is the weight of the mp-CA/GO (g); M is the molar mass of metals (g mol^{-1}); and C_i and C_f represent the initial and final concentration of metal ions in solution, respectively (mM).

3. Results and Discussion

3.1. Material Characterizations

Figure 1A illustrates the GO and PS suspensions that were used to prepare the mp-CA/GO. PS colloidal particle suspension appears milky white, whereas GO suspension appear brownish black due to high concentration. GO and PS can be dispersed in DI water and form homogeneous suspensions after a mild ultrasonic treatment. The XRD pattern of GO is presented in Figure 1D. The strong diffraction peak of GO appears at 11.5°, which can be indexed to the (002) reflections of stacked GO sheets with the interlayer distance of 0.771 nm, which allows alginate to exist between layers of GO [17–19]. The mp-CA/GO with excellent mechanical strength and structural stability was acquired by the help of GO. The morphology of PS and GO was characterized by SEM, as depicted in Figure 1B,C. The surface of the GO film has noticeable folds, which are attributed to the flexible, ultra-thin 2D lamination of GO (Figure 1C). As the most commonly used template material, PS colloidal particles have an excellent uniformity with a diameter of ca. 720 nm (Figure 1B), this uniformity is important for preparing materials with controlled porous structures.

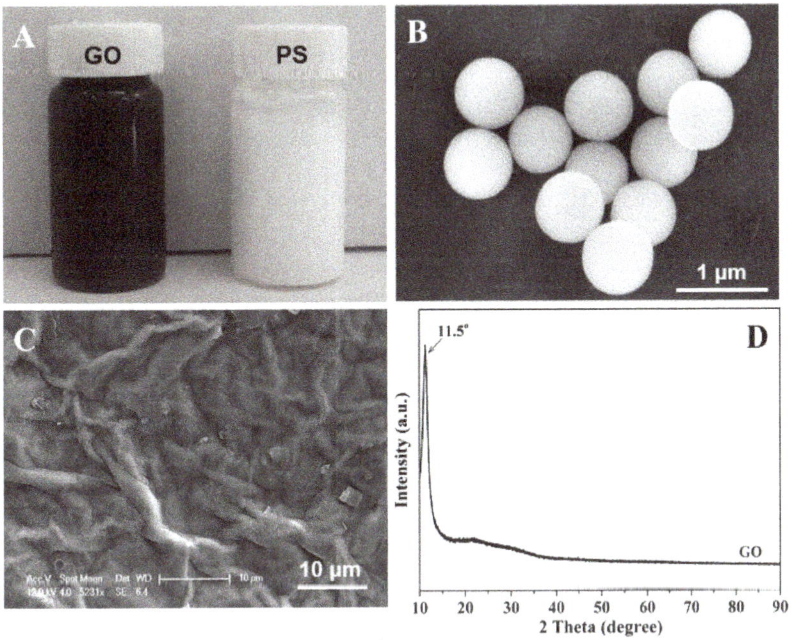

Figure 1. (**A**) Digital photo of GO and PS solutions, SEM images of (**B**) PS colloidal particles and (**C**) GO film, and (**D**) XRD pattern of GO.

The FT-IR spectra of GO, calcium alginate (CA) and mp-CA/GO are presented in Figure 2. In the three spectra, the absorption peaks around 3360 and 1412 cm^{-1} are ascribed to the O–H and –COOH, respectively. The weak peak around 2930 cm^{-1} in the spectra of CA and mp-CA/GO belongs to the asymmetric stretching vibration of C–H, in which the "egg-box" structures formed by sodium alginate macromolecule and calcium ion limit the C–H stretching [20,21]. The peaks at 3179, 1715 and 1620 cm^{-1} in the spectrum of GO are attributed to the stretching vibrations of –OH, –COOH and C=C in the sp^2 carbon skeletal network, respectively [22]. Compared with the pure GO, the peaks of mp-CA/GO downshift from 3380 to 3346 cm^{-1} and 1612 to 1599 cm^{-1}, respectively. It can be attributed to the intermolecular hydrogen bonds formed between GO sheets and alginate [23].

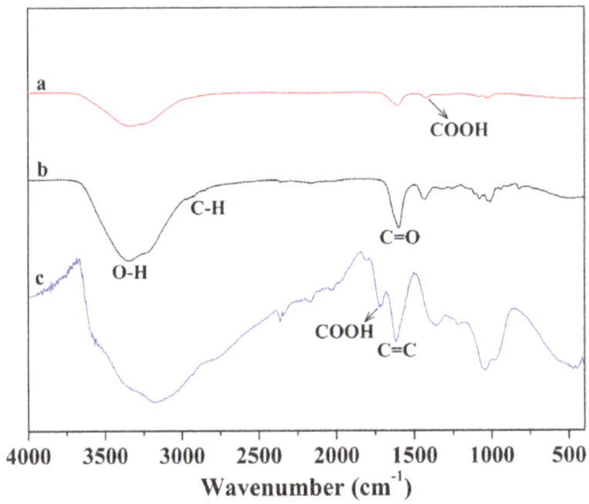

Figure 2. FT-IR spectra of (a) mp-CA/GO, (b) CA and (c) GO.

The SEM photos of the mp-CA/GO and CA are displayed in Figure 3. The PS beads with a diameter of ca. 720 nm are homogeneously dispersed in a composite aerogel sphere (Figure 3A), the spaces that were originally occupied by the PS beads remain as macropores after removing the particle template, and mp-CA/GO has a uniform diameter of approximately 720 nm macropore (Figure 3B). The surface appears more uneven in the mp-CA/GO with controlled pore structure than CA (Figure 3C), this phenomenon is beneficial to adsorb metal ions. In addition, the diameter of the pore structure can be regulated by changing the diameter of PS colloidal particle template, thus allowing us to design the required pore size of materials.

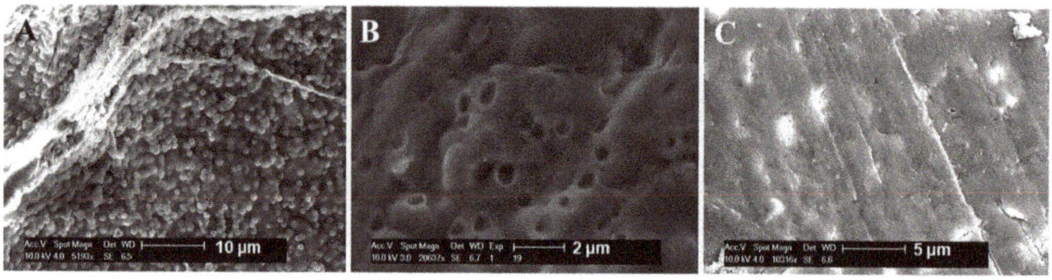

Figure 3. SEM photos of (**A**,**B**) mp-CA/GO/CA and (**C**) CA.

3.2. *Effect of pH*

In order to evaluate the effect of pH on adsorption performance, we conducted experiments at pH values that range from one to seven. As can be seen in Figure 4, adsorption capacity of the as-prepared mp-CA/GO increased significantly when the pH increased from one to three, and then remained stable with a further increase in pH. This result can be due to the change in the ionic state of the amino and carboxyl groups. When the mp-CA/GO was in strong acidic circumstance (pH < 3), functional groups (amino and carboxyl) were protonated. However, amino and carboxyl groups are deprotonated with an increase in the solution pH ($3 \leq pH \leq 7$) [10,24]. Therefore, mp-CA/GO has wide pH range to meet the needs of real application.

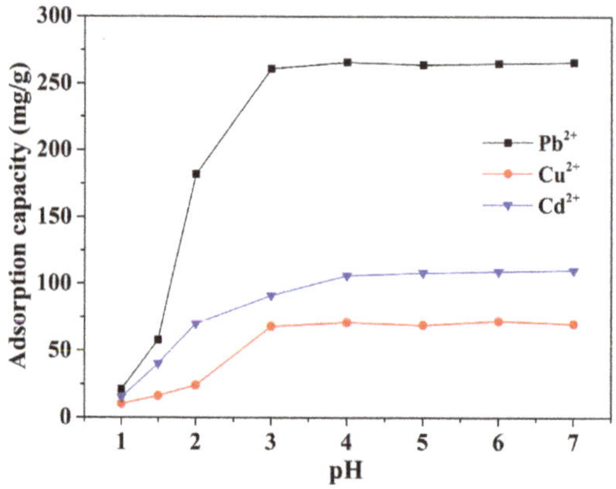

Figure 4. Effect of pH on metal ions adsorption. (~ 50 mg of mp-CA/GO was equilibrated with 50 mL of 1.5 mM Pb^{2+}, Cu^{2+}, or Cd^{2+} at 25 °C for 40 min).

3.3. *Effect of the Contact Time and Environmental Temperature*

The effect of contact time on the adsorption behavior of mp-CA/GO was also evaluated (Figure 5). The adsorption rate of Pb^{2+}, Cu^{2+}, and Cd^{2+} onto the mp-CA/GO was quite fast, thereby completing the adsorption process within 40 min. The adsorption rate was clearly high at the initial adsorption period, possibly due to the controlled porous structures and abundant vacant sites of aerogel. The diffusion rate of metal ions was accelerated through the nanoporous structure of mp-CA/GO, thereby indicating the primary objective of preparing porous structure using PS colloidal particles as template. Furthermore, pseudo-first-order and pseudo-second-order kinetic models have been employed to investigate the adsorption kinetics during the adsorption process. The results summarized in Table 1 show that the pseudo-second-order kinetic model fits well with the kinetic data according to the values of R^2, thus suggesting that chemical sorption is a rate determining step in the adsorption process.

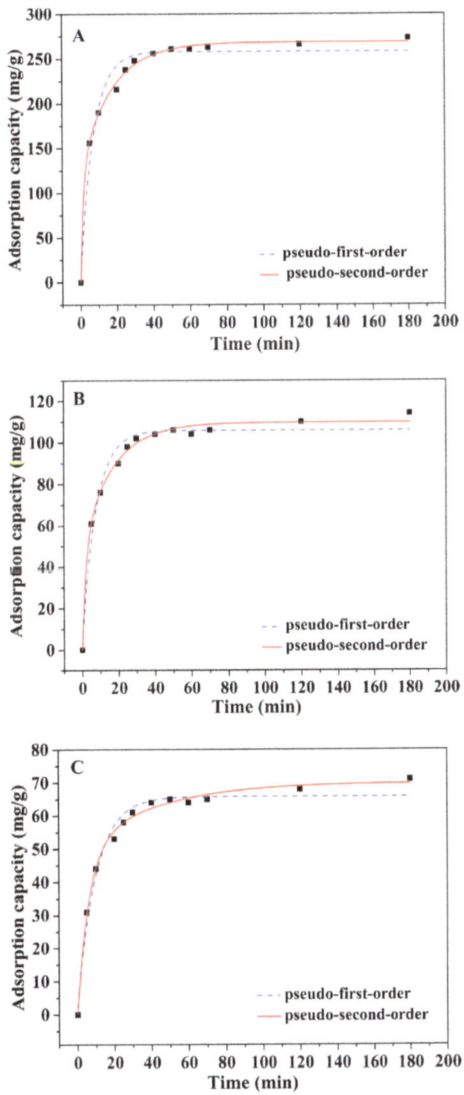

Figure 5. Adsorption kinetics studies of the (**A**) Pb^{2+}, (**B**) Cd^{2+}, and (**C**) Cu^{2+}. (~ 50 mg of mp-CA/GO was equilibrated with 50 mL of 1.5 mM Pb^{2+}, Cu^{2+}, or Cd^{2+} solution at 25 °C).

Table 1. Kinetic parameters for Pb^{2+}, Cd^{2+}, and Cu^{2+} ions adsorption on the mp-CA/GO.

Kinetic Model	Formula	Parameters	Pb^{2+}	Cu^{2+}	Cd^{2+}
pseudo-first-order	$q_t = q_e(1 - \exp(-k_1 t))$	q_e (mg g^{-1})	257.849	65.779	105.737
		k_1 (L min^{-1})	0.158	0.124	0.113
		R^2	0.962	0.975	0.968
pseudo-second-order	$q_t = q_e(1 - 1/(1 + q_e k_2 t))$	q_e (mg g^{-1})	268.284	69.985	109.589
		k_2 (L min^{-1})	0.004	0.002	0.002
		R^2	0.996	0.992	0.991

Furthermore, the effect of environmental temperature on the adsorption behavior for heavy metal ions was also investigated. As seen in Figure 6, the adsorption capacity of the mp-CA/GO increased slowly with the increase in temperature. However, its improvement in terms of adsorption performance was vague, possibly because the large specific surface area and controlled porous structure of mp-CA/GO allowed the metal ions to acquire additional binding sites and rapid ion diffusion rate on the material. The adsorbent's weak sensitivity to temperature is crucial to practical applications, thereby enabling the mp-CA/GO to be potentially applied to the practical treatment of heavy metal ions.

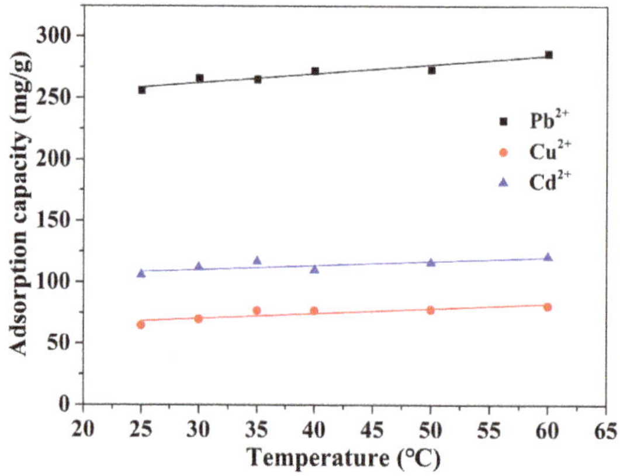

Figure 6. Effect of the environmental temperature on metal ions adsorption. (~ 50 mg of mp-CA/GO was equilibrated with 50 mL of 1.5 mM Pb^{2+}, Cu^{2+}, or Cd^{2+} solution at different temperatures for 40 min).

3.4. Maximum Adsorption Capacity of the mp-CA/GO

The maximum adsorption capacity is one of the most important features for the adsorbent. As seen in Figure 7, the adsorption capacity increases dramatically with the initial concentration from 0.1 mM to four mM for the mp-CA/GO. After exceeding four mM, the increment in mp-CA/GO has leveled off, possibly because of the lack of adequate functional groups to accommodate additional metal ions. The maximum adsorption capacities of mp-CA/GO for Pb^{2+}, Cu^{2+}, and Cd^{2+} are 368.2, 98.1 and 183.6 mg/g, respectively. These values are higher than most of the reported heavy metal ion adsorbents [25–43] (Table 2).

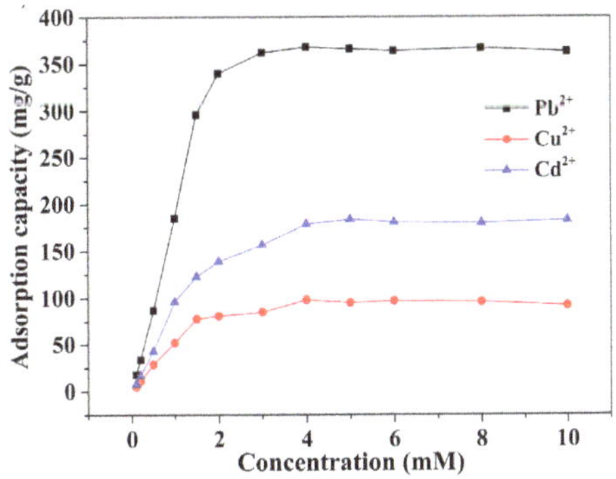

Figure 7. Effect of the initial metal ion concentration on adsorption. (~ 50 mg of mp-CA/GO was equilibrated with 50 mL of Pb^{2+}, Cu^{2+}, or Cd^{2+} solution at 25 °C for 40 min).

Table 2. Comparison of adsorption capacity of various adsorbent for Pb^{2+}, Cu^{2+}, and Cd^{2+}.

Adsorbent	Heavy Metals (Adsorbate)	Maximum Adsorption Capacity (mg/g)	Year of Publication	Reference
Amino functionalized mesoporous silica	Pb^{2+}, Ni^{2+}, Cd^{2+}	57.7 (Pb^{2+}), 12.4 (Ni^{2+}), 18.3 (Cd^{2+})	2009	[25]
Nano-alumina	Pb^{2+}, Cr^{3+}, Cd^{2+}	100.0 (Pb^{2+}), 100.0 (Cr^{3+}), 83.3 (Cd^{2+})	2010	[26]
Amino functionalized magnetic graphenes composite	Pb^{2+}, Hg^{2+}, Cr^{6+}, Cd^{2+}	28.0 (Pb^{2+}), 23.0 (Hg^{2+}), 17.3 (Cr^{6+}), 27.8 (Cd^{2+})	2014	[27]
Polydopamine microspheres	Pb^{2+}	165.8	2017	[28]
Polyving alcohol/polyacrylic acid double network gel	Pb^{2+}, Cd^{2+}	195.0 (Pb^{2+}), 115.9 (Cd^{2+})	2015	[29]
Biochar-alginate capsule	Pb^{2+}	263.2	2013	[30]
Polyaniline/calcium alginate composite	Pb^{2+}, Cu^{2+}	357.0 (Pb^{2+}), 79.0 (Cu^{2+})	2012	[31]
Silica modified calcium alginate-xanthan gum hybrid bead composite	Pb^{2+}	18.9	2013	[32]
Activated carbon-calcium alginate composite	Pb^{2+}	15.7	2016	[33]
Alginate-SBA-15 composite	Pb^{2+}	222.2	2013	[34]
Soy protein hollow microspheres	Pb^{2+}, Zn^{2+}, Cr^{3+}, Cd^{2+}, Cu^{2+}, Ni^{2+}	235.6 (Pb^{2+}), 255.0 (Zn^{2+}), 52.9 (Cr^{3+}), 120.8 (Cd^{2+}), 115.0 (Cu^{2+}), 177.1 (Ni^{2+})	2013	[35]
Magnetic alginate beads	Pb^{2+}	50	2012	[36]
γ-Fe_2O_3 nanoparticles	Pb^{2+}, Cu^{2+}	69.0 (Pb^{2+}), 34.0 (Cu^{2+})	2017	[37]
Magnetic chitosan/graphene oxide imprinted Pb^{2+}	Pb^{2+}	79.0	2016	[37]
Chitosan coated calcium alginate	Pb^{2+}	106.9	2016	[39]
Hydroxyapatite/chitosan porous material	Pb^{2+}	264.4	2015	[40]
Calcite-poly(ethyleneimine) nanostructured rod	Pb^{2+}	240	2013	[41]
Nanostructured vaterite-poly(ethyleneimine) hybrid	Pb^{2+}	2762	2014	[42]
Alginate-melamine hybrid	Pb^{2+}	287.7	2018	[43]
mp-CA/GO	Pb^{2+}, Cu^{2+}, Cd^{2+}	368.2 (Pb^{2+}), 98.1 (Cu^{2+}), 183.6 (Cd^{2+})	This work	This work

To study the adsorption behavior and predict which adsorption system is favorable. Langmuir adsorption isotherm and Freundlich adsorption isotherm models were applied in this study (Figure 8). The isotherm constants were calculated from the experimental data and are presented in Table 3.

The results show that the Langmuir isotherm appears to be a favorable model to supervise the adsorption process, which indicates the adsorption process was a monolayer adsorption.

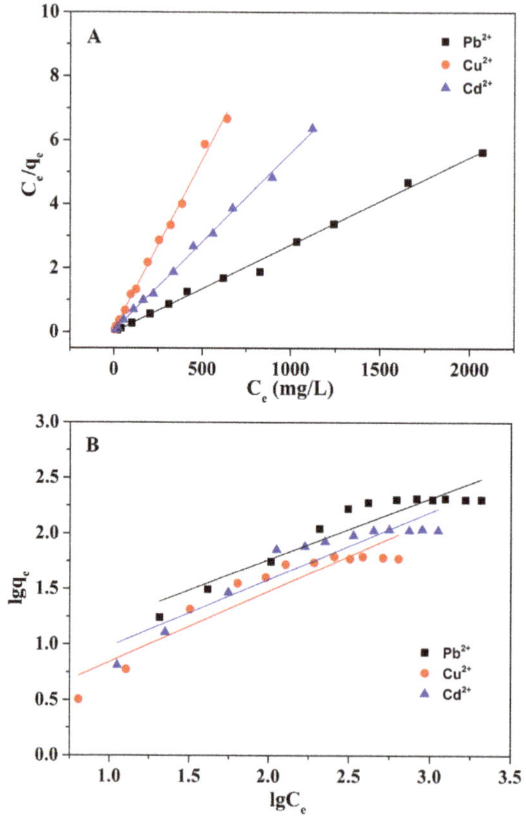

Figure 8. (**A**) Langmuir and (**B**) Freundlich adsorption isotherms studies of the Pb^{2+}, Cd^{2+}, and Cu^{2+}. (~ 50 mg of mp-CA/GO was equilibrated with 50 mL of Pb^{2+}, Cu^{2+}, or Cd^{2+} solution at 25 °C for 40 min).

Table 3. Isotherm parameters for Pb^{2+}, Cd^{2+}, and Cu^{2+} ions adsorption on the mp-CA/GO.

Isotherm Model	Formula	Parameters	Pb^{2+}	Cu^{2+}	Cd^{2+}
Langmuir	$C/q = C/q_e + 1/(q_e b)$	q_e (mg/g)	366.835	180.274	96.693
		b (L/mg)	0.493	0.473	0.415
		R^2	0.994	0.998	0.995
Freundlich	$lgq = lgK + 1/nlgC$	K (L/mg)	5.263	3.863	3.573
		n	2.163	1.663	1.862
		R^2	0.878	0.873	0.864

3.5. Adsorption Mechanism

In order to clarify the adsorption mechanism of the mp-CA/GO, we monitored the amount of each metal ion in the adsorption process. It can be seen from Figure 9 that the mole amount of Pb^{2+}, Cu^{2+} and Cd^{2+} in the solution decreased while Ca^{2+} increased as the reaction proceeded. This illustrates that, in the first adsorption process, the adsorption of heavy metal ions is mainly completed by ion exchange. However, the mole amount of the adsorbed Pb^{2+}, Cu^{2+}, or Cd^{2+} is higher than that of the desorbed Ca^{2+} during the different adsorption periods, indicating that the ion exchange is only one of the mechanisms. In order to fully understand the essence of adsorption, we further analyzed the O one s spectra before and after Pb^{2+}, Cu^{2+}, and Cd^{2+} adsorption (Figure 10). The two peaks at 530.88 and 529.767 eV in the O one s spectrum are attributed to the oxygen-containing functional groups. New peaks appeared at 530.461, 530.728, and 530.866 eV, respectively, after the adsorption of Pb^{2+}, Cu^{2+}, and Cd^{2+}, indicating that the oxygen groups of on mp-CA/GO were involved in chemisorption of Pb^{2+}, Cu^{2+}, and Cd^{2+}. Therefore, the excellent adsorption performance of mp-CA/GO is attributed to the combined action of chemical coordination and ion exchange.

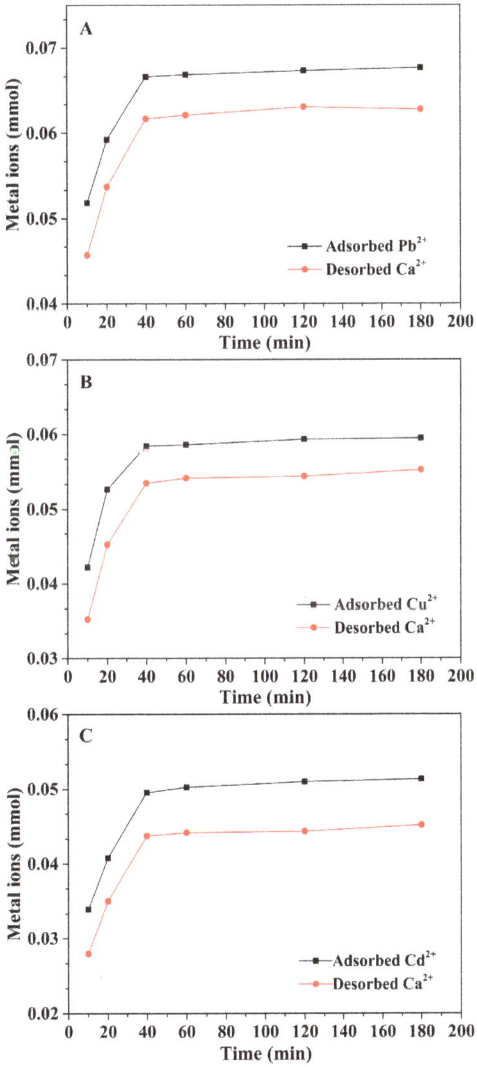

Figure 9. The mole amount of (**A**) Pb^{2+}, Ca^{2+}, (**B**) Cu^{2+}, Ca^{2+} and (**C**) Cd^{2+}, Ca^{2+} in different adsorption times.

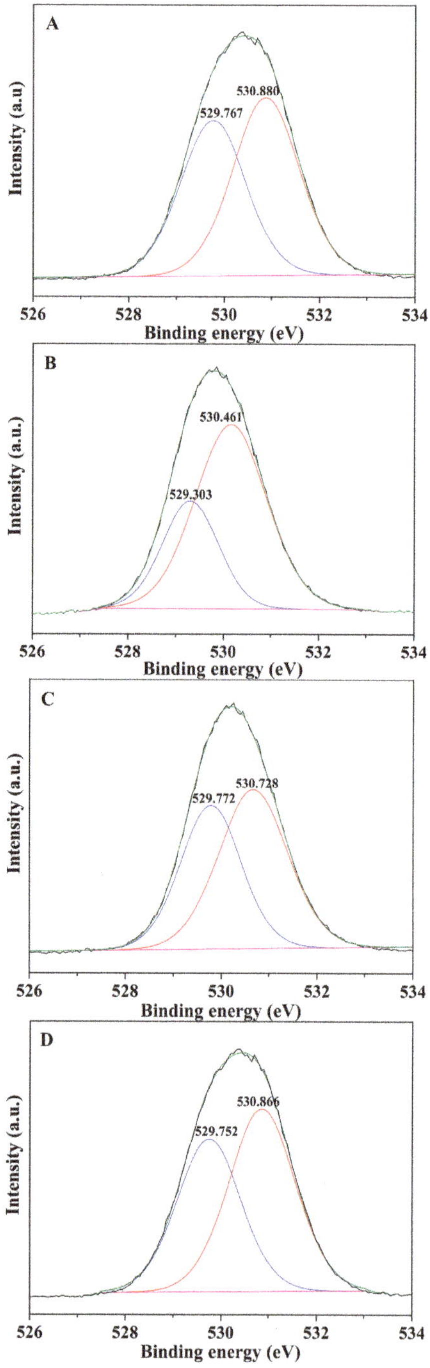

Figure 10. O1s XPS spectra of (**A**) mp-CA/GO, (**B**) mp-CA/GO with Pb^{2+} adsorption, (**C**) mp-CA/GO with Cu^{2+} adsorption, and (**D**) mp-CA/GO with Cd^{2+} adsorption.

3.6. Regeneration Research

In the regeneration process, the mp-CA/GOs that were individually loaded with Pb^{2+}, Cu^{2+}, and Cd^{2+} were immersed into 50 mL of 0.07 M HNO_3 solution. Then, the mp-CA/GO was separated and washed sequentially with DI water, calcium hydroxide solution, and then DI water till the final eluent to neutral.

The recyclability of mp-CA/GO is depicted in Figure 11. The mp-CA/GO could remove Pb^{2+}, Cu^{2+} and Cd^{2+} after 20 adsorption-desorption cycles with a performance loss within five %, possibly attributed to the GO, as reinforcing fillers, to maintain the mechanical strength and elasticity of mp-CA/GO during the continuous adsorption-desorption process.

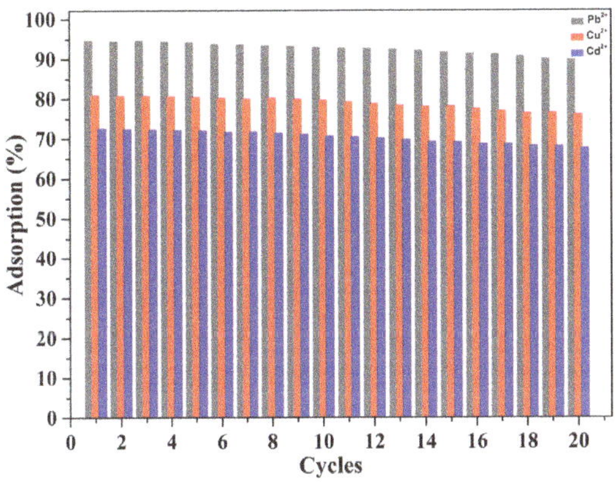

Figure 11. Regeneration research of mp-CA/GO. (~ 50 mg of mp-CA/GO and 1.5 mM of Pb^{2+} solution were used here).

4. Conclusions

In this study, the mp-CA/GO with controlled pore structure was used as an efficient solid adsorbent to remove heavy metal ion in wastewater. The fabrication of a controlled porous structure has two advantages. First, the porous structure increases the surface area of the material itself, thereby enabling the adsorbent material to provide additional binding sites for heavy metal ions. The removal rates for Pb^{2+}, Cu^{2+}, and Cd^{2+} were 95.4%, 81.2%, and 73.2%, respectively. Second, existing porous structure significantly accelerates the diffusion rate of ions in the adsorption process, thus enabling the adsorption process to complete within 40 min. Furthermore, the mp-CA/GO could be regenerated through a simple acid washing process and used repeatedly with little loss in performance. Additionally, the wide pH application range and the weak sensitivity to temperature also allowed the mp-CA/GO to be potentially applied to actual heavy metal sewage treatment.

Author Contributions: Z.W. and L.P. conceived and designed the experiments; L.P. performed the experiments; L.P. and Q.Y. analyzed the data; Z.W. and L.P. wrote the paper; R.H. made revisions of the manuscript.

Acknowledgments: We gratefully acknowledge the national natural science foundation of China (21407004), Anhui province office of education (KJ2017A345) and Program for Innovative Research Team in Anqing Normal University for financial support.

Conflicts of Interest: The authors declare no conflict of interest.

References

1. Volesky, B. Detoxification of metal-bearing effluents: Biosorption for the next century. *Hydrometallurgy* **2001**, *59*, 203–216. [CrossRef]
2. Voisin, H.; Bergstrom, L.; Liu, P.; Mathew, A.P. Nanocellulose-based materials for water purification. *Nanomaterials* **2017**, *7*, 57. [CrossRef] [PubMed]

3. Babel, S.; Kurniawan, T.A. Low-cost adsorbents for heavy metals uptake from contaminated water: A review. *J. Hazard. Mater.* **2003**, *97*, 219–243. [CrossRef]
4. Wang, Z.Q.; Wu, A.G.; Ciacchi, L.C.; Wei, G. Recent advances in nanoporous membranes for water purification. *Nanomaterials* **2018**, *8*, 65. [CrossRef] [PubMed]
5. Cataldo, S.; Lazzara, G.; Massaro, M.; Muratore, N.; Pettignano, A.; Riela, S. Functionalized halloysite nanotubes for enhanced removal of lead(II) ions from aqueous solutions. *Appl. Clay Sci.* **2018**, *156*, 87–95. [CrossRef]
6. Hepel, M.; Dentrone, L. Controlled incorporation of heavy metals from aqueous solutions and their electrorelease using composite polypyrrole films. *Electroanalysis* **1996**, *8*, 996–1005. [CrossRef]
7. Hepel, M.; Zhang, X.M.; Stephenson, R.; Perkins, S. Use of electrochemical quartz crystal microbalance technique to track electrochemically assisted removal of heavy metals from aqueous solutions by cation-exchange composite polypyrrole-modified electrodes. *Microchem. J.* **1997**, *56*, 79–92. [CrossRef]
8. Cavallaro, G.; Gianguzza, A.; Lazzara, G.; Milioto, S.; Piazzese, D. Alginate gel beads filled with halloysite nanotubes. *Appl. Clay Sci.* **2013**, *72*, 132–137. [CrossRef]
9. Deze, E.G.; Papageorgiou, S.K.; Favvas, E.P.; Katsaros, F.K. Porous alginate aerogel beads for effective and rapid heavy metal sorption from aqueous solutions: Effect of porosity in Cu^{2+} and Cd^{2+} ion sorption. *Chem. Eng. J.* **2012**, *209*, 537–546. [CrossRef]
10. Wang, Z.Q.; Huang, Y.G.; Wang, M.; Wu, G.H.; Geng, T.M.; Zhao, Y.G.; Wu, A.G. Macroporous calcium alginate aerogel as sorbent for Pb^{2+} removal from water media. *J. Environ. Chem. Eng.* **2016**, *4*, 3185–3192. [CrossRef]
11. Wang, Z.Q.; Jin, P.X.; Wang, M.; Wu, G.H.; Sun, J.Y.; Zhang, Y.J.; Dong, C.; Wu, A.G. Highly efficient removal of toxic Pb^{2+} from wastewater by an alginate-chitosan hybrid adsorbent. *J. Chem. Technol. Biot.* **2018**, *93*, 2691–2700. [CrossRef]
12. Lagoa, R.; Rodrigues, J.R. Kinetic analysis of metal uptake by dry and gel alginate particles. *Biochem. Eng. J.* **2009**, *46*, 320–326. [CrossRef]
13. Jiao, C.L.; Xiong, P.G.; Tao, J.; Xu, S.J.; Zhang, D.S.; Lin, H.; Chen, Y.Y. Sodium alginate/graphene oxide aerogel with enhanced strength-toughness and its heavy metal adsorption study. *Int. J. Biol. Macromol.* **2016**, *83*, 133–141. [CrossRef] [PubMed]
14. Yang, X.Z.; Zhou, T.Z.; Ren, B.Z.; Hursthouse, A.; Zhang, Y.Z. Removal of Mn(II) by sodium alginate/graphene oxide composite double-network hydrogel beads from aqueous solutions. *Sci. Rep.* **2018**, *8*, 10717. [CrossRef] [PubMed]
15. Lui, G.; Li, G.; Wang, X.L.; Jiang, G.P.; Lin, E.; Fowler, M.; Yu, A.P.; Chen, Z.W. Flexible, three-dimensional ordered macroporous TiO_2 electrode with enhanced electrode electrolyte interaction in high-power li-ion batteries. *Nano Energy* **2016**, *24*, 72–77. [CrossRef]
16. Choi, B.G.; Yang, M.; Hong, W.H.; Choi, J.W.; Huh, Y.S. 3D macroporous graphene frameworks for supercapacitors with high energy and power densities. *Acs Nano* **2012**, *6*, 4020–4028. [CrossRef] [PubMed]
17. Kim, D.Y.; Kim, M.; Kim, D.W.; Suk, J.; Park, J.J.; Park, O.O.; Kang, Y. Graphene paper with controlled pore structure for high-performance cathodes in LiO_2 batteries. *Carbon* **2016**, *100*, 265–272. [CrossRef]
18. Song, N.J.; Chen, C.M.; Lu, C.X.; Liu, Z.; Kong, Q.Q.; Cai, R. Thermally reduced graphene oxide films as flexible lateral heat spreaders. *J. Mater. Chem. A* **2014**, *2*, 16563–16568. [CrossRef]
19. Xiao, F.; Yang, S.X.; Zhang, Z.Y.; Liu, H.F.; Xiao, J.W.; Wan, L.; Luo, J.; Wang, S.; Liu, Y.Q. Scalable synthesis of freestanding sandwich-structured graphene/polyaniline/graphene nanocomposite paper for flexible all-solid-state supercapacitor. *Sci. Rep.* **2015**, *5*, 9359. [CrossRef] [PubMed]
20. He, Y.Q.; Zhang, N.N.; Gong, Q.J.; Qiu, H.X.; Wang, W.; Liu, Y.; Gao, J.P. Alginate/graphene oxide fibers with enhanced mechanical strength prepared by wet spinning. *Carbohydr. Polym.* **2012**, *88*, 1100–1108. [CrossRef]
21. Maria, H.; Dustin, B.; Matthew, M.; Magdalena, S.; Kaitlin, C. Assembly of Gold Nanoparticles Induced by Metal Ions. In *Functional Nanoparticles for Bioanalysis, Nanomedicine, and Bioelectronic Devices*; Hepel, M., Zhong, C.J., Eds.; ACS Symposium Series; Oxford University Press: Oxford, UK, 2012; Volume 1, pp. 207–240.
22. Braccini, I.; Perez, S. Molecular basis of Ca^{2+}-induced gelation in alginates and pectins: The egg-box model revisited. *Biomacromolecules* **2001**, *2*, 1089–1096. [CrossRef] [PubMed]
23. Wan, Y.Z.; Chen, X.Q.; Xiong, G.Y.; Guo, R.S.; Luo, H.L. Synthesis and characterization of three-dimensional porous graphene oxide/sodium alginate scaffolds with enhanced mechanical properties. *Mater. Express* **2014**, *4*, 429–434. [CrossRef]

24. Huang, Y.G.; Wang, Z.Q. Preparation of composite aerogels based on sodium alginate, and its application in removal of Pb^{2+} and Cu^{2+} from water. *Int. J. Biol. Macromol.* **2018**, *107*, 741–747. [CrossRef] [PubMed]
25. Heidari, A.; Younesi, H.; Mehraban, Z. Removal of Ni(II), Cd(II), and Pb(II) from a ternary aqueous solution by amino functionalized mesoporous and nano mesoporous silica. *Chem. Eng. J.* **2009**, *153*, 70–79. [CrossRef]
26. Afkhami, A.; Saber-Tehrani, M.; Bagheri, H. Simultaneous removal of heavy-metal ions in wastewater samples using nano-alumina modified with 2,4-dinitrophenylhydrazine. *J. Hazard. Mater.* **2010**, *181*, 836–844. [CrossRef] [PubMed]
27. Guo, X.Y.; Du, B.; Wei, Q.; Yang, J.; Hu, L.H.; Yan, L.G.; Xu, W.Y. Synthesis of amino functionalized magnetic graphenes composite material and its application to remove Cr(VI), Pb(II), Hg(II), Cd(II) and Ni(II) from contaminated water. *J. Hazard. Mater.* **2014**, *278*, 211–220. [CrossRef] [PubMed]
28. Zhang, Q.R.; Yang, Q.G.; Phanlavong, P.; Li, Y.X.; Wang, Z.K.; Jiao, T.F.; Peng, Q.M. Highly efficient lead(II) sequestration using size-controllable polydopamine microspheres with superior application capability and rapid capture. *ACS Sustain. Chem. Eng.* **2017**, *5*, 4161–4170. [CrossRef]
29. Chu, L.; Liu, C.B.; Zhou, G.Y.; Xu, R.; Tang, Y.H.; Zeng, Z.B.; Luo, S.L. A double network gel as low cost and easy recycle adsorbent: Highly efficient removal of Cd(II) and Pb(II) pollutants from wastewater. *J. Hazard. Mater.* **2015**, *300*, 153–160. [CrossRef] [PubMed]
30. Do, X.H.; Lee, B.K. Removal of Pb^{2+} using a biochar-alginate capsule in aqueous solution and capsule regeneration. *J. Environ. Manage.* **2013**, *131*, 375–382. [CrossRef] [PubMed]
31. Jiang, N.N.; Xu, Y.T.; Dai, Y.Q.; Luo, W.A.; Dai, L.Z. Polyaniline nanofibers assembled on alginate microsphere for Cu^{2+} and Pb^{2+} uptake. *J. Hazard. Mater.* **2012**, *215*, 17–24. [CrossRef] [PubMed]
32. Zhang, S.; Xu, F.; Wang, Y.F.; Zhang, W.Z.; Peng, X.L.; Pepe, F. Silica modified calcium alginate-xanthan gum hybrid bead composites for the removal and recovery of pb(ii) from aqueous solution. *Chem. Eng. J.* **2013**, *234*, 33–42. [CrossRef]
33. Cataldo, S.; Gianguzza, A.; Milea, D.; Muratore, N.; Pettignano, A. Pb(II) adsorption by a novel activated carbon—alginate composite material. A kinetic and equilibrium study. *Int. J. Biol. Macromol.* **2016**, *92*, 769–778. [CrossRef] [PubMed]
34. Cheraghali, R.; Tavakoli, H.; Sepehrian, H. Preparation, characterization and lead sorption performance of alginate-SBA-15 composite as a novel adsorbent. *Sci. Iran.* **2013**, *20*, 1028–1034.
35. Liu, D.G.; Li, Z.H.; Li, W.; Zhong, Z.R.; Xu, J.Q.; Ren, J.J.; Ma, Z.S. Adsorption behavior of heavy metal ions from aqueous solution by soy protein hollow microspheres. *Ind. Eng. Chem. Res.* **2013**, *52*, 11036–11044. [CrossRef]
36. Idris, A.; Ismail, N.S.M.; Hassan, N.; Misran, E.; Ngomsik, A.F. Synthesis of magnetic alginate beads based on maghemite nanoparticles for Pb(II) removal in aqueous solution. *J. Ind. Eng. Chem.* **2012**, *18*, 1582–1589. [CrossRef]
37. Rajput, S.; Singh, L.P.; Pittman, C.U.; Mohan, D. Lead (Pb^{2+}) and copper (Cu^{2+}) remediation from water using superparamagnetic maghemite (γ-Fe_2O_3) nanoparticles synthesized by flame spray pyrolysis. *J. Colloid Interface Sci.* **2017**, *492*, 176–190. [CrossRef] [PubMed]
38. Wang, Y.H.; Li, L.L.; Luo, C.N.; Wang, X.J.; Duan, H.M. Removal of Pb^{2+} from water environment using a novel magnetic chitosan/graphene oxide imprinted Pb^{2+}. *Int. J. Biol. Macromol.* **2016**, *86*, 505–511. [CrossRef] [PubMed]
39. Mousa, N.E.; Simonescu, C.M.; Patescu, R.E.; Onose, C.; Tardei, C.; Culita, D.C.; Oprea, O.; Patroi, D.; Lavric, V. Pb^{2+} removal from aqueous synthetic solutions by calcium alginate and chitosan coated calcium alginate. *React. Funct. Polym.* **2016**, *109*, 137–150. [CrossRef]
40. Lei, Y.; Guan, J.J.; Chen, W.; Ke, Q.F.; Zhang, C.Q.; Guo, Y.P. Fabrication of hydroxyapatite/chitosan porous materials for Pb(II) removal from aqueous solution. *RSC Adv.* **2015**, *5*, 25462–25470. [CrossRef]
41. Marzo, A.M.L.; Pons, J.; Merkoci, A. Multifunctional system based on hybrid nanostructured rod formation, for sensoremoval applications of Pb^{2+} as a model toxic metal. *J. Mater. Chem. A* **2013**, *1*, 13532–13541.
42. Lopez-Marzo, A.M.; Pons, J.; Merkoci, A. Extremely fast and high Pb^{2+} removal capacity using a nanostructured hybrid material. *J. Mater. Chem. A* **2014**, *2*, 8766–8772. [CrossRef]
43. Li, K.T.; Wu, G.H.; Wang, M.; Zhou, X.H.; Wang, Z.Q. Efficient removal of lead ions from water by a low-cost alginate-melamine hybrid sorbent. *Appl. Sci.* **2018**, *8*, 1518. [CrossRef]

© 2018 by the authors. Licensee MDPI, Basel, Switzerland. This article is an open access article distributed under the terms and conditions of the Creative Commons Attribution (CC BY) license (http://creativecommons.org/licenses/by/4.0/).

nanomaterials

MDPI

Review

The Roles of Nanomaterials in Conventional and Emerging Technologies for Heavy Metal Removal: A State-of-the-Art Review

Mahesan Naidu Subramaniam, Pei Sean Goh *, Woei Jye Lau and Ahmad Fauzi Ismail *

Advanced Membrane Technology Research Centre, School of Chemical and Energy Engineering, Faculty of Engineering, Universiti Teknologi Malaysia, Johor 81310, Malaysia; mahesannaidu@yahoo.com (M.N.S.); wjlau@petroleum.utm.my (W.J.L.)

* Correspondence: peisean@petroleum.utm.my (P.S.G.); afauzi@utm.my (A.F.I.)

Received: 15 March 2019; Accepted: 12 April 2019; Published: 17 April 2019

Abstract: Heavy metal (HM) pollution in waterways is a serious threat towards global water security, as high dosages of HM poisoning can significantly harm all living organisms. Researchers have developed promising methods to isolate, separate, or reduce these HMs from water bodies to overcome this. This includes techniques, such as adsorption, photocatalysis, and membrane removal. Nanomaterials play an integral role in all of these remediation techniques. Nanomaterials of different shapes have been atomically designed via various synthesis techniques, such as hydrothermal, wet chemical synthesis, and so on to develop unique nanomaterials with exceptional properties, including high surface area and porosity, modified surface charge, increment in active sites, enhanced photocatalytic efficiency, and improved HM removal selectivity. In this work, a comprehensive review on the role that nanomaterials play in removing HM from waterways. The unique characteristics of the nanomaterials, synthesis technique, and removal principles are presented. A detailed visualisation of HM removal performances and the mechanisms behind this improvement is also detailed. Finally, the future directions for the development of nanomaterials are highlighted.

Keywords: heavy metal removal; nanomaterials; adsorption; photocatalysis; membrane

1. Introduction

In a bid to bolster its economic growth, heavy industrialization across developing countries is rampant. One hindsight with this phenomenon is the creation of many types of poorly managed waste that eventually seep deep into the environment via air, soil, and water. The damage of natural resources takes place as the direct consequence of the release of hazardous substances [1]. In particular, water is the prominent recipient of many types of pollutant. The continuous pollution of water, in tandem with the increased demand for fresh water due to industrialization and population growth, has strained water resource to a breaking point. Various types of pollutants, such as natural organic matters (NOM), oil, pathogens, and heavy metals (HM) have badly afflicted water. HM pollutions has been a prominent issue, as the HM infusing into the water sources can be produced from various human activities, including mining, agriculture, and electronic industries [2]. It is imperative that HM contents in waterways are controlled efficiency in view of their poisonous and toxic nature towards all organisms. In a broad context, HMs are a group of trace metals with an atomic density of less than 5 ± 1 g/cm^3 [3]. Cadmium, manganese, iron, arsenic, and mercury are some of the prominent HMs that are commonly found in wastewater [4]. These metals usually exist in the form of ions in waterways and soil. They can pose a health hazard to both humans and the ecosystem via avenues, such as direct ingestion or in contact with contaminated water or soil, drinking water that has been contaminated with HM, ingestion of foodstuff laden with HM (plants or aquatic life), as well as the accumulation of HM via the food chain. Critically, organisms are unable to metabolize and excrete HMs out of the body. The presence of HM ions in water samples imparts great ecological impact due to its toxicity and bioaccumulation, as aquatic life is known to accumulate significant concentration of metals in water where the presence of such metal in water samples are below the detection levels [5].

The sources of HM ions in the environment can be focused on two origins, i.e., natural sources and human activities. The former source includes phenomena, such as landslides, weathering, and volcanic eruption. These activities have significantly contributed to HM pollution, as these events release trapped HM ions into the environment. Anthropogenic activities are also a major source of HM pollution. Activities, such as mining, smelting operations, industrial production and manufacturing industry, and agricultural use of HM in the form of pesticides are major causes of HM pollution on soil and in water. Controlling the leaching of HM into the environment is important. Various legislations and laws have been enacted to control the number of pollutants that are released by industry players into the environment. Table 1 shows the maximum level of HM content in water samples across different local and global agencies.

Table 1. Maximum level of heavy metal (HM) content in water samples. Table was reproduced from [6], with permission from EDP Sciences, 2017.

Agency	Permissible Level of HM (mg/g)								
	Cd (II)	Cr (III)	Co (II)	Cu (II)	Pb (II)	Fe (II)	Mn (II)	Hg (II)	Ni (II)
National Agency for Food and Drug Administration and Control (NAFDAC)	0.0	NM	NM	NM	0.0	NM	NM	0.0	NM
United States Environmental Protection Agency (USEPA)	0.005	0.1	0.1	1.3	0.015	0.3	0.05	0.002	0.1
World Health Organization (WHO)	0.003	NM	NM	0.01	0.01	0.3	0.4	0.001	0.07
Department of Environment (DOE), Malaysia	0.005	0.05	NM	1.0	0.1	1.0	0.2	NM	NM

Remark: NM refers to 'not mentioned'.

On the other hand, the remediation of these HMs is also essential in reducing the impact of HM on human health. As of now, numerous strategies have been developed to separate these HMs from water sources. Remediation techniques include the employment of absorbents, coagulation of HM ions, chemical precipitation, membrane filtration, electrodialysis, and photocatalysis. Each of these techniques has its unique advantages as well as disadvantages that are particularly associated to their efficiency for large scale HM remediation. Unique nanomaterials that were developed by researchers are at the center of these treatment methods. The discovery and the subsequent extensive research on nanomaterials developed in different geometries have brought about different physicochemical properties that cannot be expressed by bulk materials. Nanomaterials, such as carbon nanotubes (CNT) [7], graphene [8], titania nanotubes (TNT) [9], hybrid metal-non-metal nanomaterials, such as graphite silica [10] and graphene oxide-magnetite nanocomposite [11], and other metal oxide-based nanomaterials have been extensively used to remove HM ions from polluted waters, with a varying degree of success. These nanomaterials can be advantageously used to remove HM ions via adsorption, owing to their large adsorption capacities. Especially, nanomaterials of different geometries have exhibited high HM removal efficiency due to unique properties, such as large surface area, specific surface charge values, surface functionality, and HM ions binding capabilities. These nanomaterials can also be incorporated with other treatment techniques, such as adsorptive membranes and composite membranes to work synergistically and to further improve HM ion removal efficiency.

Over the years, a number of comprehensive reviews have been made in regard to nanomaterials and their applications for environmental remediation. Khin et al. compiled a comprehensive review of the application of nanomaterials as a viable solution in the removal of various types of pollutants and biological contaminants [12]. Azzouz et al. published a more recent review on the utilization of nanomaterials as sorbents in a solid-phase pollutant extraction system for environmental samples [13]. The author discussed the use of various types of nanomaterials as potential sorbents for analytical applications. Jeevanandam et al. prepared a review on the history, sources, toxicity, and regulations of nanoparticles and nanostructured materials [14]. Despite the efforts that have been made in this area, very small number of reviews has been made to discuss the significant role of nanomaterials in removing pollutants from wastewater [15]. It is worth providing insights regarding the functionality of nanomaterials of different compositions, characteristics and structures, as well as their key roles in the removal of HM ions, particularly in different remediation techniques from various sources.

In this paper, an overview of prominent sources and types of HM containments in the water sources is first provided. Next, various nanomaterial synthesis techniques were used to prepare unique nanomaterials especially for HM ions removal from water sources are discussed. In the main body of this review, the performances of nanomaterials in various HM removal strategies are evaluated. Finally, the current hurdles and future directions of the HM removal strategies that are based on nanomaterials are highlighted.

2. Heavy Metal Ions in the Environment

2.1. Sources of Heavy Metal

HMs have great ecological consideration, due to their toxicity and accumulation. Fish might accumulate significant concentrations of metals in water in which those metals are below the limit of detection in a routine water sampling and analysis [16]. The sources of HM in the ecosystem are focused on human activities and natural phenomena. Some of the sources include the usage of HM-laden pesticides and naturally occurring HM from the Earth's core via activities, such as soil erosion and volcanic eruption.

Pesticides (ethylene dibromide, and methyl bromide fungicides), insecticides (dithiocarbamates, and captan), and herbicides (paraquat, diquat, and 2, 4-dichlorophenoxyacetic acid) have been used for many decades as means to improve the survival of vegetations that are planted for commercial purpose and used to kill off various kinds of pest that harm the quality of crops. HMs are usually used as active compounds in the aversion of pest. Copper, usually in the form of copper sulfate and mercuric chloride, is used for its anti-fungicidal properties. Sodium, in the form of sodium dichromate works as a cotton defoliant [17]. Zinc phosphide is used as a rodenticide, whist cadmium chloride is used as a fungicide [18]. These pesticides laden with HM are usually prepared in ionic form, which later dissociates after being dispersed into the soil. Their nature allows for them to reside in water and travel far away from the point of origin. Pesticides can be taken up by plants, dissociate into soil, or carried away by residual water into other water bodies in contact, such as rivers and lakes [19]. Absorption of pesticides by plants removes them from the environment. However, plants cannot metabolize these compounds and stay in the plants, which can be transferred to other organisms that consume these plants, such as animals and humans itself [20]. Evidence of HM ending up in plant specimens is aplenty [21–23]. Via phytoextraction, plants can absorb HMs, which are essential in plant growth [24]. However, they also absorb HMs, such as cadmium, chromium, and lead, which do not serve them in any biological function. Consequently, the bioaccumulated HM will be passed along the food chain [25].

2.2. Effect of Heavy Metal

2.2.1. Effect of Heavy Metal Ions Towards the Environment

The largest contributor of HM in the air is the usage of hydrocarbons, such as gasoline, diesel, and petrol. HM, such as arsenic, lead, and cadmium, are emitted when these hydrocarbons are combusted [26]. Volcanic eruptions produce hazardous impacts to the environment, as the deterioration of social and chemical conditions and the gases (carbon dioxide, sulfur dioxide, carbon monoxide, and hydrogen sulfide) that are released during eruptions, various organic compounds and HMs, such as mercury, lead, and gold, are also released [4]. HMs enter plant and animal tissues via air inhalation, diet, and manual handling. Motor vehicle emissions are a major source of airborne contaminants, including arsenic, cadmium, cobalt, and nickel. HMs leaching from industrial and consumer waste can pollute water sources (groundwater, lakes, streams, and rivers); acid rain can exacerbate this process by releasing HMs that are trapped in soils. However, the risk of these metals entering the food chain is highly dependent on the mobility of the metal cations and its bioavailability in soil. The metal cations are bound to negatively charged particles in soil, such as clay and organic matter. When these metal cations detach from the negatively charged particles, they become available to be absorbed by plants and other organisms that live in the soil [25]. Plants are exposed to HMs through the uptake of water, and are stored until animals, which then transfer the HM into the animal's body, consume these plants. The ingestion of plant and animal-based foods that are laden with HM is one of the worrying sources of HMs in humans. The presence of such inorganic pesticide can also degrade the soil due to the accumulation of compounds at undesirable levels [27]. Absorption through skin contact with soil is another potential source of HM contamination. Studies have also shown that HMs can be accumulated in the plant tissues of *Sebera acuminate* and *Thlaspi caerulescens*, as they cannot be metabolized [21]. Arsenic poisoning is one of the most prevalent HM cases across the globe, which usually occurs by drinking water that is contaminated with arsenical pesticides, natural mineral deposits, or inappropriate disposal of arsenical chemicals. A work done by Sim et al. revealed that rivers in Sarawak have experienced extensive land use and logging, and hence suffer from contaminations of HMs, such as arsenic, chromium, and copper [28]. Another similar research also revealed that River Pra and its tributaries displayed an enrichment in HM ions [2]. Figure 1 shows the route of absorption, distribution, and excretion that are related to the exposure of HMs and inorganic pesticides [29].

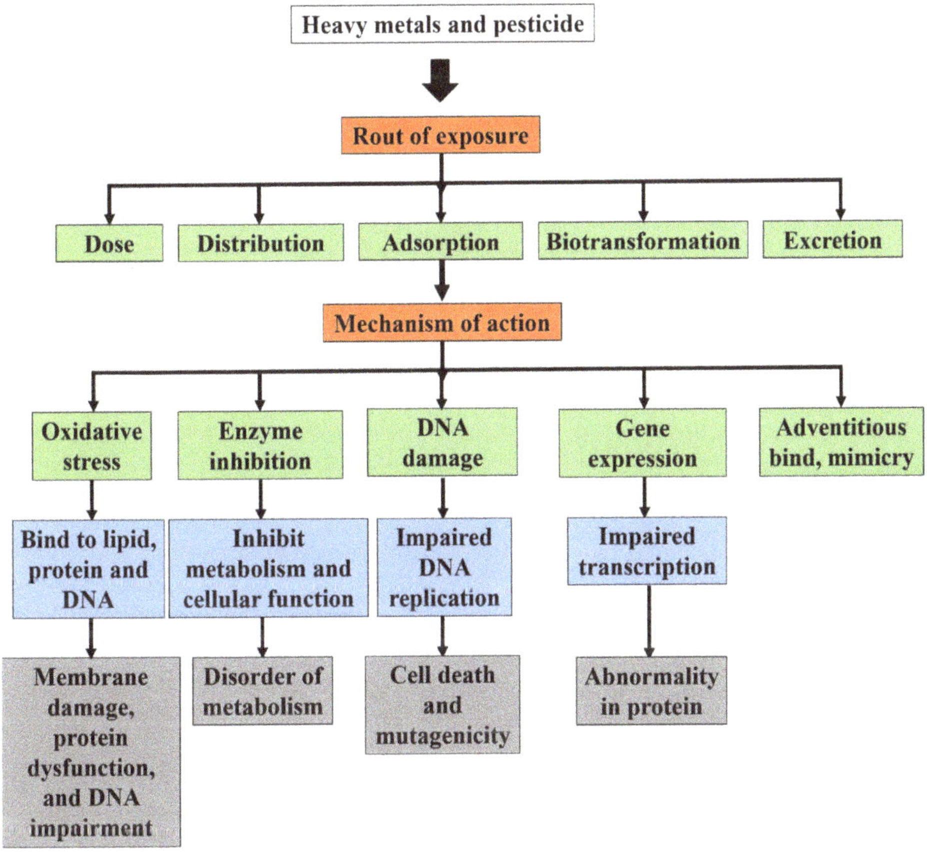

Figure 1. Common route of absorption, distribution, and excretion related to the exposure of HMs and inorganic pesticides. Adapted from [29], with permission from Frontiers, 2017.

2.2.2. Effect of Heavy Metal Ions Towards Humans

Human can be afflicted with HM poisoning due to the consumption of food or water that is laden with HM. They stay in the human body system and result in constant accumulation of different types of HM, ince humans are not able to metabolize it. Exposure to As (V) leads to an accumulation of As (V) in tissues, such as skin, hair, and nails, resulting in various clinical symptoms, such as hyperpigmentation and keratosis. There is also an increased risk of skin, internal organ, and lung cancers. Lead is known to disrupt the balance between the production of free radicals and the generation of antioxidants to detoxify the reactive intermediates or to repair the resulting damage [30]. Reactive oxidation species (ROS) may cause structural damage to cells, proteins, nucleic acid, membranes, and lipids, resulting in a stressed situation at cellular at very high concentrations [31]. Lead is also known to disrupt biological processes, such as cell adhesion, intra- and inter-cellular signaling, protein folding, maturation, apoptosis, ionic transportation, enzyme regulation, and release of neurotransmitters [32]. Aluminium is a common HM that is used in the production of carbonated drink cans and cooking utensils [33]. The WHO postulated that aluminium exposure is probably a risk factor in the onset of Alzheimer disease in humans, whist reports have also suggested that humans can also be afflicted with contact dermatitis and irritant dermatitis [30,34]. Mercury, as a type of HM that is commonly found in many types of seafood and being previously used in dental amalgam, can also potentially harm humans due to their acute toxicology. Mercury tends to be tightly bound in the brain, spinal cord, ganglia, autonomic ganglia, and peripheral motor neurons upon entering into the human body [35,36].

2.3. *The Chemistry of Heavy Metal Ions*

HM ions tend to present in salt or oxide forms. The HM ions ionic values are generally between +2 to +6, which indicates that metals placed in group 2 to group 6 of the periodic table of elements are categorised as HMs. Differing ionic states of HMs can significantly affect the suitability of different HM removal techniques and their conditions or parameters. For instance, the ionic state of HM can be altered in different pH conditions, which in turn affects the electrostatic interaction between the nanoadsorbent and the HM ions [37]. For the removal of HM via adsorption, the pH condition between 4 and 7 is favourable, as it greatly improves surface coordination, electrostatic attraction, and co-precipitation, which will result a higher removal rate. Increment from pH 4 to 7 reduces protonation phenomena, as it increases the H^+ availability. As the pH increases, the overall charge positivity increases and it also increases the interaction between the HM and adsorbent. This also increases the formation of chelate complexes between metal cations and the lone pair of electrons on the sulfur and nitrogen atom [38]. HM ions, such as Pb (II) and Hg (II) are efficiently adsorbed by neutrally charged adsorbents in the pH range between 4 and 7. On the other hand, when the absorbent surface is charged, the pH can also be manipulated to attain a parameter where the greatest electrostatic interaction between adsorbent and HM occur. Besides, the surface functional group also plays a role in the adsorption capacity. Al-Senani et al. exhibited that the removal of Co (II) and Cd (II) was effective above pH 9, exhibiting a removal rate of more than 98% [39]. This is due to the functional group exhibited by the adsorbent, which includes carboxyl, hydroxyl, and amine, all being involved in the binding mechanism between adsorbent and HM.

3. Nanomaterials-Assisted Approach for Heavy Metal Removal

The removal of HM can be carried out by several traditional techniques. Some of the methods include electrochemical treatment, photocatalysis, coagulation, adsorption, chemical precipitation, and membrane technologies, such as reverse osmosis and nanofiltration. However, in this review paper, three main HM removal technologies that have portrayed the greatest potential for large scale application have been focused upon. The HM removal techniques in focus are adsorption, photocatalytic reduction and membrane filtration/adsorption.

3.1. *Adsorption of Heavy Metals*

Adsorption is a process where a surface holds a molecule onto it. Adsorption happens via two phenomena, which are physisorption or chemisorption [40]. IUPAC defines adsorption as the increase in concentration of a dissolved substance at the interface of a condensed and a liquid phase due to the operation of surface forces [41]. Physisorption occurs when forces, such as intermolecular forces, are used to attach the absorbate onto the absorbent meanwhile chemisorption involves valence forces of the same kind as those operating in the formation of chemical compounds [42]. For any adsorption process, kinetic and isotherm studies are performed to evaluate the adsorption phenomena, rate, and efficiency [43]. Some of the common parameters that affect adsorption include the pH of the aqueous solution, the interaction between the adsorbent and adsorbate, surface charge, the surface area of adsorbate, and the size of adsorbent and adsorbate. Adsorption is a viable way of removing HM from water bodies. Many researchers have conducted a study investigating the ideal pH value, adsorbate surface area, and porosity and surface charge for effective adsorption of various types of HM ions [44]. Figure 2 illustrates the possible interaction of HM ions with the surface if adsorbents, such as polyaniline/TiO_2 composites [45], cation exchanged porous zeolite [46], and binary metal adsorption by biochar derived from activated sludge [47].

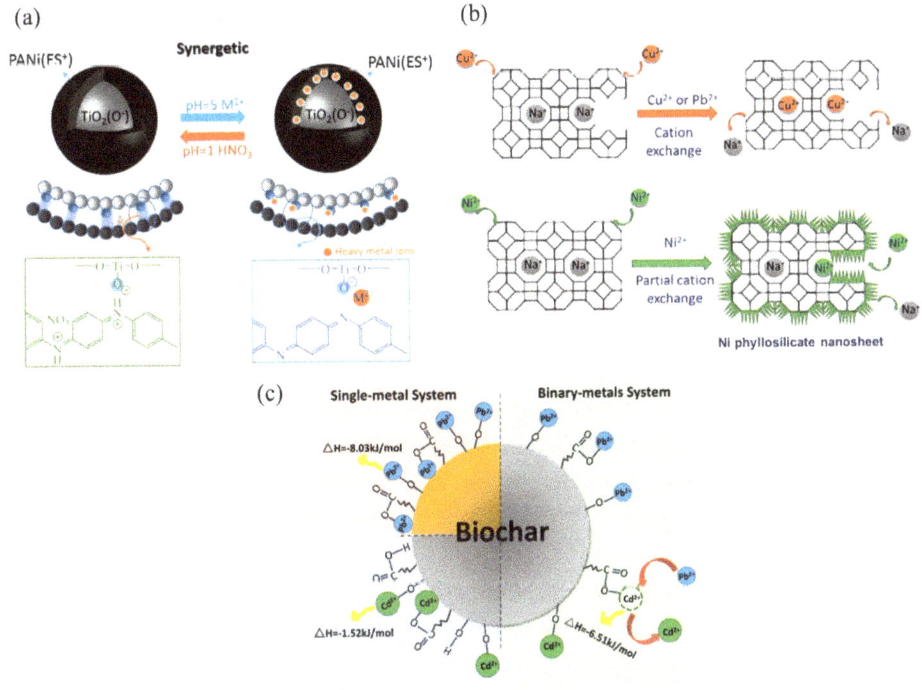

Figure 2. Schematic illustration of the adsorption of HM via the surface of (**a**) hybrid polyaniline/TiO_2 nanocomposite adsorbents. Adapted from [45], with permission from Elsevier, 2018. (**b**) cation exchange by hierarchically porous zeolite for improved adsorption of cationic HMs. Adapted from [46], with permission from Elsevier, 2019. and (**c**) selective HM ion adsorption by biochar in a single and binary metal system. Adapted from [47], with permission from Elsevier, 2019

3.2. Photocatalytic Reduction of Heavy Metal

Among many types of wastewater remediation techniques that have been discovered by researchers, photocatalysis remains one of the best methods, simply because it is able to destroy or reduce the pollutant, rather than just mitigate, trap, or isolate them. The application of various types of semiconductor materials, including titanium dioxide (TiO_2) and zinc oxide (ZnO) as a light responsive material to treat wastewater that is laden with organic have garnered much of the attention of emerging researchers. These semiconductors can produce strong oxidative free radicals that are capable of destroying a large range of organic pollutant and reduce HM ions when it is irradiated with light sources. These semiconductors harvests the photons and excite electrons into a higher energy state when light is irradiated, producing electron pair holes that are transferred on the surface of the semiconductor, which in turn, produces these ROS such as OH● and $O_2^{\bullet -}$. Figure 3 shows a brief illustration on the excitation of an electron in a structure of photocatalyst and the subsequent creation of ROS species.

Many types of research have been focused on the fine-tuning and the modification of these semiconductors for efficient degradation of organic pollutants since semiconductors are able to efficiently degrade NOM. The possibility of using photocatalyst in the removal of HM is less explored, as it is impossible to degrade metallic ions. However, photocatalyst has shown promise as a means to reduce the harmful effects of HM by reducing the metal ions into less harmful by-products. The reduction of HM ions is a viable means for the treatment of HM pollution. Cr (VI) is significantly harmful to organisms, even in small trace amount, as compared to Cr (II). Hence, the common practice for remediation of Cr (VI) is reducing it to Cr (II). This is where photocatalysts can play a prominent role in HM waste remediation. Based on the current research trend, the employment of photocatalyst has been commonly used as means to reduce Cr (VI), which can be abundantly found in contaminated water, sourced from industries, such as electroplating, pigmentations, and so on [48]. The photoreduction of Cr (VI) is elucidated in Equation (1) and Equation (2) [49];

$$Cr_2O_7^{2-} + 14H^+ + 6e^- \rightarrow 2Cr^{3+} + 7H_2O \tag{1}$$

$$2H_2O + 2h^+ \rightarrow H_2O_2 + 2H^+ \tag{2}$$

Cr (VI) has been extensively researched, because the hexavalent equivalent is much more poisonous than its divalent variant, Cr (III) counterpart. When the photocatalyst is irradiated with photons, they absorb it and excite an electron towards the particle surface. The Cr (VI) ions consume these electrons, allowing single photoreduction. This, in turn, makes the catalyst reactive due to the continuous presence of an electron-hole pair. To exploit this mechanism, research regarding photoreduction of Cr (VI) metal ions are paired with an organic pollutant, as the organic pollutant becomes an electron source for the electron deprived photocatalyst, which initiates a chain reaction on continuous photoreduction of Cr (VI) and the degradation of an organic pollutant. This mechanism provides further proof that photocatalysis can work for the photodegradation and photoreduction of wastewater mixed with HM ions and organic pollutants, hence enhancing its versatility [49].

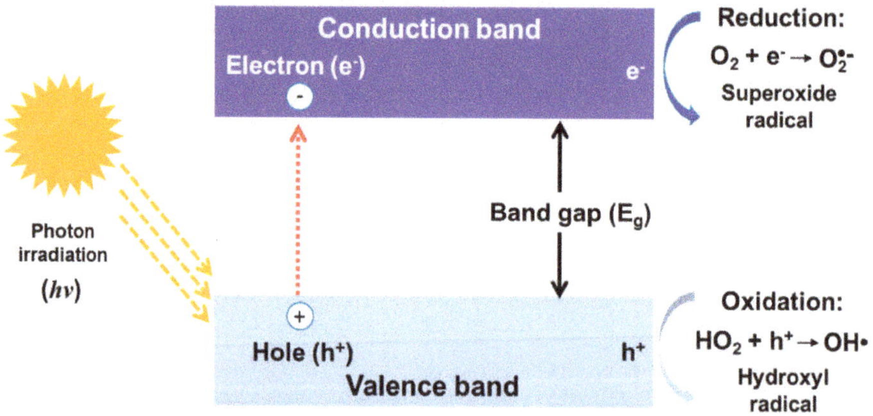

Figure 3. Excitation of an electron in a structure of photocatalyst and subsequent creation of ROS. Adapted from [48], with permission from Elsevier, 2018.

3.3. Membrane Filtration/Adsorption

Membrane separation is another emerging technology that has shown great promise in HM separation from polluted waters. The focus for development of a membrane-based solution for HM remediation intensified when researchers discovered problems in other common remediation, which include the poor reusability of adsorbents, high cost of material development, and difficulty in separating nanomaterials from the water after remediation [50]. The employment of membrane-based remediation method can address these problems, as membrane that is incorporated with novel nanomaterials is a one-step method that can be reused while maintaining excellent rejection of HM in aqueous solution without the need of chemicals or pre-treatments. In the search for the development of membranes that are low in cost, high reusability, greater selectivity, better water transport, and high HM ions rejection, researchers are exploring the possibility of incorporation of various types of nanomaterials in a bid to impart the unique characteristics of nanomaterials into membranes. The addition of high surface area nanomaterials has developed highly adsorptive membranes [51], while the incorporation of highly hydrophilic nanomaterials on the membrane surface has significantly improved the water permeability of membranes [52]. Photocatalytic hybrid nanomaterials, such as Graphitic carbon nitride (g-C_3N_4) quantum dots (QD) [53] and TNT array [54], are also developed to allow for simultaneous photocatalysis and membrane filtration in efficient wastewater treatment [55]. Figure 4 shows the utilization of several membrane-based technologies in the remediation of water bodies laden with HM ions [56–58].

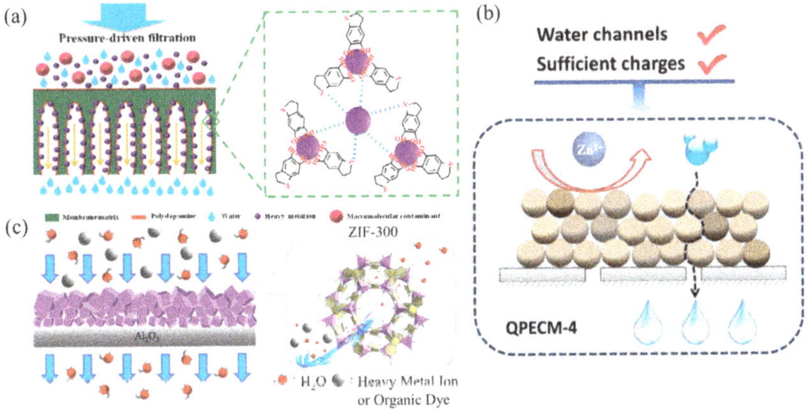

Figure 4. HM removal via (**a**) adsorptive membrane technique. Adapted from [56], with permission from Elsevier, 2017. (**b**) surface-charged modified membrane repellent. Adapted from [57], with permission from Elsevier, 2019. and (**c**) size exclusion of HM ions. Adapted from [58], with permission from Elsevier, 2019.

Even though there are many ways that membranes are employed to remove HM, including surface charged membranes for HM ion repellent, membrane distillation [59], adsorptive membrane [60], size exclusion removal [57], and more, there are two popular ways where HM can be removed from water bodies using membrane-based remediation, which is size exclusion removal, or using adsorptive membranes [61]. Commonly, membrane works by sieving the molecules according to size. Only particles that are larger than the pore size are retained. In this case, the removal of HM requires the employment of nanofiltration (NF) membranes, as the pore size of common ultrafiltration (UF) membranes are large and would allow HM ions to pass through. However, NF membranes face one prominent problem. They will restrict the movement of water through the membrane since the pore size is very small, which drastically reduces membrane flux [62]. To overcome this, researchers have explored the idea of adding novel nanomaterials into these membranes to improve the permeation, whilst maintaining or enhancing the HM rejection rate. Nanomaterials, such as halloysite nanotubes [63], CNT [64], metal oxide nanoparticles [65], and many more are blended into a polymeric matrix or are deposited on the membrane surface to improve selectivity and permeation, whilst maintaining or enhancing rejection capabilities. The improved performance of membranes is owed to the modification of characteristics due to the presence of nanomaterials. Materials, such as CNT's, create a pathway for water transport through the membrane matrix, increasing water permeation [66]. Other nanomaterials, such as amine grafted SiO_2, are able to impart superhydrophilicity into a polymeric membrane, which is generally hydrophobic in nature [67]. Adsorptive filtration is another way where membranes can be employed for HM removal. Albeit, adsorption itself is a promising way to treat HM waters, it is bugged by a few problems. Sorbents always tend to float on the water surface, and this would render proper mixing with water to achieve maximum contact between adsorbent and HM ions untenable [68]. If the adsorbents are properly mixed, then their form as a particle would require a form of method to separate the used adsorbent from the water source [69]. This would incur additional time and resource. Hence, adsorptive membranes have been developed to overcome this, where potent adsorbents are immobilized into the membrane matrix [70]. Adsorptive membranes exhibit the ability to trap the HM ions and at the same time, allow the filterability of water, and produce clean permeates with the metal trapped in the membrane matrix. Commonly, UF membranes are used as an adsorptive membrane, as it has a larger pore size as compared to the NF membranes. To overcome the trade-off for the low rejection of HM ions, nano-adsorbent is incorporated into UF membranes to maintain both high membrane flux due to the larger pore size and excellent HM removal.

4. Nanomaterials for Heavy Metal Removal

Adsorption has shown the greatest potential in terms of cost and effectiveness, even though all of the techniques are able to perform admirably in removing HM from water sources, while membrane filtration techniques have also shown good performance with long term stability and usage [62]. In addition to this, photocatalysis, have shown significant promise, owing to its non-selective degradation, excellent ability to mineralise pollutants, and good reusability [71]. Nanomaterials have played very significant roles in advancing the HM removal technologies in all of the methods that were mentioned above. Nanomaterials are described as a material with a length of 1 to 100 nm in at least one dimension [14]. Their small size allows for them to exhibit unique properties that are not shown in bulk, with some examples of properties including increased surface pore and surface area [72], improved electrical properties, increased material strength and conductivity, and self-cleaning properties [73]. Nanomaterials can be classified into metallic, non-metallic, and their composites. Metal oxides and semiconductors are common sources of metallic nanomaterials, whereas non-metallic nanomaterials include carbonous nanomaterials, such as (CNT) [74] and graphene. Nanocomposites can be multiphase nanomaterials, where one part is defined as a nanomaterial in terms of size, whilst the other may also be a nanomaterial or they can be materials that are larger with bulk-type property. In addition to this, nanomaterials are also known according to the shape that they exhibit. Some of the nanomaterial shapes include nanoparticles [75] (spherical/globular), nanosheets [76], nanowires, nanoflowers [77], nanotubes [78], and nanorods [79].

4.1. *Motivation of Using Nanomaterial for Heavy Metal Removal*

The utilization of nanoparticles for the remediation of environmental problems has shown remarkable potential in line with the rapid development of nanoscience and nanotechnology. The modification in atomic level to produce particles that are independent in nanoscale has provided a myriad of novel characteristics that cannot be found in bulk materials. Nanomaterials can be synthesized using bottom-up approach to carefully tailor the desired properties, such as surface charges and functionalities to interact with HM. Nanostructured materials have shown exceptionally high surface area and porosity, higher efficiency as an absorbent due to their superior surface to volume ratio, improved solubility, abundant reaction sides, photocatalytic properties, great surface charge, and lighter in weight or mass. These nanomaterials can also be modified via various techniques, such as surface grafting and gamma irradiation, to increase its surface reactivity [80]. CNT has also been used as a prominent adsorbent for HM removal due to their high surface area to volume ratio and their highly tunable characteristics. Various functional groups, such as hydroxyl and carbonyl groups, which can provide new adsorption sites, are easily tuned on the surface of CNT [81]. Clay can also be utilized as an efficient absorbent of HM via nanoscience. Studies have shown that common clay that is combined with activated carbon, another low cost and a common material, are able to absorb HM ions, such as Cd (II), Ba (II), and Cu (II) from pulp wastewater [82]. The synthesized absorbent was able to exhibit a surface area of close to 800 m^2/g, which is ten-fold larger when compared to individual nanomaterial. Another research also showed that the creation of nanomaterial via a nanocasting process using mesoporous hybrid material with ZnO and TiO_2 exhibited a surface area between 120–332 m^2/L [83]. The absorbent also showed that it could be reused up to three times due to the micrometer-sized structure with high surface area, which has the benefit of reducing the overall cost in the adsorption process. In addition to this, semiconducting nanomaterials, which are also known as photocatalyst, can be employed in HM reduction. This is based on the fact that it has good optical properties and the energy band can be easily modified through facile modification or hybrid to render improved properties, such as lower bang gap energy, lower recombination rates, and larger active sites for photocatalysis. Nanomaterials of unique features can also be incorporated with other technologies to create a synergistic improvement in HM removal. For instance, the incorporation of novel nanomaterials into polymeric membrane matrices, such as CNT, TiO_2, and hydrous manganese dioxide (HMO) can increase the pathways for water transport, impart photocatalytic activity, and increase membrane hydrophilicity respectively. These nanomaterials can also deposit onto the membrane surface that can greatly govern the selectivity of membranes, something that is not possible with polymers alone. The utilization of nanoscience also enabled the creation of functional nanomaterials from waste source, such as the formation of a biogenic iron (Fe) compound at a size of 500 nm, using a natural microbial consortium that was sourced from an abandoned mine containing iron oxides (Fe_2O_3) and siderite by bioreduction of ferric citrate [84].

4.2. Classification of Nanomaterials

As researchers further pushed the boundary of nanoscience in the development of novel, functional materials, they discovered that the shape of nanomaterials could be manipulated. The differing shapes opened a vast array of new and unique characteristics that were not possible in its benign shape. Carbonous materials led the way, where carbon was used as the building block in building two different variants, namely CNT and graphene [76]. CNT follows the shape of tubes, while graphene took the shape of sheets, in multiple layers. Other forms also emerged, which includes the formation of nanorods and nanoflowers of different, as shown in the micrographs in Figure 5 [85–88].

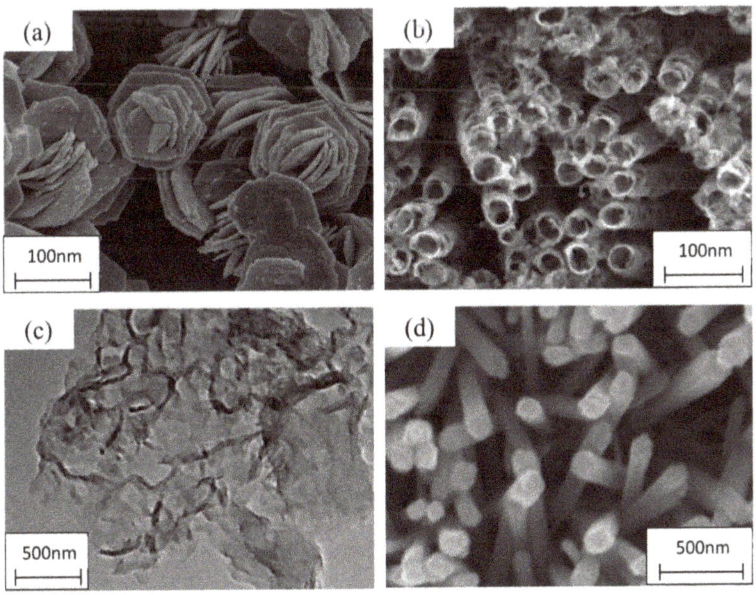

Figure 5. Examples of nanomaterial structures (**a**) nanoflowers. Adapted from [85]. (**b**) nanotubes. Adapted from [86]. (**c**) nanosheets. Adapted from [87], and (**d**) nanorods. Adapted from [88].

The change in structure can bring about new and unique features that otherwise are not exhibited in the bulk phase. CNT's are produced when sheets of carbon, which are called graphene, are rolled up to produce single-walled CNT's or multi-walled CNT's [89]. The formation of tubular structure allows graphene to be 400 times stronger than steel, allowing for conducting electricity and even working as a semiconductor, something that is not possible in bulk carbon [90]. This allows CNT's to be cheaper and more environmentally stable materials in the development of electrical and electronic products that rely heavily on rare earth metals. TiO_2 is a metal oxide that can be used in paints and personal cosmetic products as pigments due to its stable and environmentally safe nature. TNT have been developed and it has shown great promise in gas sensing and increased photocatalytic activity [91]. The development of tubular and porous structure, such as nanotubes and hydrated manganese oxide nanoparticles, have shown great improvement in terms of effective surface area that can reach the region of 400 m^2/g, which is a ten-fold increase [92]. Higher surface area leads to increased reaction sites, which is valuable in the catalytic industry. Nanoflowers have been developed from materials, such as TiO_2 and Fe_2O_3, exhibiting a superb volume to area ratio, better charge transfer, carrier immobility, and an enhanced number of adsorption sites [93]. All of these characteristics can significantly contribute to the field, such as drug delivery, catalytic process, chelation, and adsorption of HM ions [85]. In addition to these structures, nanorods/nanowires have also attracted the attention of researchers. Particularly, gold nanorods have received extensive attention, owing to their extremely attractive applications in biomedical technologies, plasmon-enhanced spectroscopies, pollutant remediation, and optical and optoelectronic devices. Copper nanoflowers have been shown to exhibit impressive adsorption of Pb (II) in aqueous solution, owing to its porous, high surface area structure, which significantly increases the availability of active sides and the presence of carboxylic (COOH) functional groups [94]. The flower structure is useful in drug delivery, as efficient drug deliveries require a carrier that is non-hydrolysable, controlled release pattern, and reducing drug toxicity. Nanoflowers tick all of these boxes, as displayed by the efficiency of sodium alginate/chitosan nanoflowers in drug delivery. Table 2 shows the types of nanomaterials and the important features that they exhibit as compared to the bulk material.

Table 2. Types of nanomaterials and the important features they exhibit as compared to bulk material.

Material	Classification	Unique Feature	Synthesis Technique	Application	Reference
Streptavidin (SA)-horseradish peroxidase (HRP)	Nanoflowers	Improved biocompatibility and attachment of the protein	Wet chemical synthesis	Biomarker detection	[95]
TiO_2 and diatomite	Nanoparticle	High surface area, improved absorbability	Wet chemical precipitation	Photocatalyst	[96]
Fe_3C	Nanoparticle	Good heating ability in magnetic fields	Hydrothermal and sonication	Magnetic hyperthermia	[97]
$BiOBr/Ti_3C_2$	Nanoparticle	Surface functionalisation	Self-assembly	HM photoreduction	[98]
ZnO	Nanorods	Improved electrode performance	Hydrothermal and sputtering	Energy nanogenerators	[99]
Au	Nanorods	Huge electric field enhancements	Direct growth and fabrication	Plasmonic spectroscopies	[100]
Ag	Nanorods	Increased dispersion and stability	Wet chemical synthesis	Transparent heaters	[101]
TiO_2	Nanotubes	High hydrophilicity, surface area	Hydrothermal	Membrane filler	[102]
SiO_2-Ge	Nanotubes	Excellent thermal transport, large surface-to-volume ratio.	-	Phonon transport	[103]
Carbon	Nanotubes	High surface area and adsorption capacity	-	Adsorption of diazinon	[104]
$Cu_3(PO_4)_2$.	Nanoflowers	Spherical, porous and hierarchical structure	Wet chemical synthesis	Photodegradation of phenol	[105]
WS_2	Nanoflowers	Increased reaction sites	Hydrothermal and reduction	Hydrogen generation	[106]

There are two ways where nanomaterials are structured, which can use a top-down or and bottom-up approach [107]. The top-down approach uses synthesis methods, such as lithography etching and exfoliating. This method is sparsely used due to the lack of versatility. Commonly, the bottom-up method is used, as it allows for control of the structure of nanomaterials at an atomic level, as different parameters can be manipulated to design the nanomaterial structure according to desire. Some of the common bottom-up methods in the design of nanomaterials include wet chemical precipitation, sol-gel, chemical vapor deposition, hydrothermal, sputtering, template growth, electrospinning, and atomic layer deposition [108]. In general, all of the techniques mentioned exhibit similar characteristics, which include high control towards to growth of structure, an abundant selection of precursors/starting material, control towards heat and temperature, high purity, and uniformity [109]. Section 4.3 greatly discusses the synthesis of nanomaterials.

4.2.1. Metallic Nanomaterials

Metallic nanomaterials are formed from metal sources, such as titanium, iron, silver, gold, manganese, copper, and many more [110]. These nanomaterials can be produced in pure metallic form or in the metal oxide forms. Metal oxides are more favourable, as they exhibit increased stability over pure metal nanomaterials. Research that was conducted on the synthesis of nanomaterials using metals is aplenty based on current literature [111–114]. Fe_2O_3 is one of the commonly used metal oxide nanoparticles. Iron can work as an adsorbent for wastewater laden with HM ions. Castro et al. produced a biogenic iron compound using metal compounds that were sourced from mining wastewater via the bioreduction of ferric citrate [84]. The biogenic nature of iron compounds has high specific surface areas and high binding energies hence act as efficient adsorbents for HMs. In addition, the bacterial matrix surrounding the iron nanoprecipitates can bind harmful metals. The metal exhibited a surface area of 56.978 m^2/g and a pore size of 8.304 nm. The large pore size allows access towards the more reactive sides to improve binding and capture the HM ions. Gold nanorods have been extensively studied due to the fact that they exhibit excellent plasmon-enhanced spectroscopies and optical and optoelectronic applications, which are hugely beneficial in the detection of HM in water samples via the colorimetric detection technique [63,77,115]. Many researchers have continuously studied upon its characteristics and its application for environmental remediation ever since the discovery of TiO_2's ability to split water [116]. TiO_2 in the form of anatase crystallinity exhibits great photocatalytic activity under ultraviolet (UV) light irradiation. The good crystallinity, together with its low band gap value (3.2 eV) and stable recombination rate, makes it an excellent candidate to reduce HMs into less harmful configurations (Cr (VI) to Cr (III)) and degrade various organic pollutants [117].

4.2.2. Non-Metallic Nanomaterials

The development of quantum dots is currently on the rise due to their superior features, such as edge morphology and increased in surface functional groups. Research that was done by Abdelsalam on the development of graphene quantum dots (GQDs) showed improved surface chemical functionality and well-defined edges, which is better when compared to the sheet structure that was exhibited by graphene [118]. GQD's adsorption capability of hydrated Cd (II) and Pb (II) calculated via density functional theory (DFT) showed that it was able to absorb the hydrated HMs through different positions and interactions, including edge and surface adsorption, interaction with unsaturated carbon atoms, and adsorption on the edge of the functionalized group. The edge of surface adsorption interaction is non-existent with other nanomaterials, which shows the uniqueness of GQDs [118]. Another example of efficient non-metallic adsorbent produced from waste is the development of CNT in the adsorption of Cd (II), Cu (II), Pb (II), and Hg (II) via chemical vapor deposition (CVD). CNT developed exhibited increased surface area and porosity, with evidence of abundant functional groups, such as O-H, C-H, C=O, C-N, C=N, phenols, aromatic rings, and aromatic groups, which provides high potential for the adsorption of HMs [47]. The developed CNT was able to act as an absorbent also shows promise, as the presence of a various functional reactive group, together with the porous structure and large network of functional group, allows it to perform well as an adsorbent for HMs, such as Cu (II), Cd (II), and Pb (II) [119].

4.2.3. Hybrid Nanomaterials

Hybrid nanomaterials can include only metallic materials and a mixture between metallic and non-metallic. Hybrid nanomaterials are strongly pursued due to the fact that these hybrid nanomaterials are able to exhibit improved, synergistic, or new properties that are not found in singular nanomaterials, metallic or non-metallic. Sharma et al. synthesized a metallic hybrid consisting of ZnO and TiO_2 hybrid monolith absorbents via nanocasting, followed with calcination at 400 °C for 5 h. The hybrid nanomaterial proved to be a better adsorbent when compared to ZiO or TiO_2 individually, as it improved the adsorption of Pb (II) and Cd (II) by more than 50% [83]. A hybrid material was developed with different ratios of clay, activated carbon and zeolite showed promising results for the adsorption of three different types of HM (Cd (II), Ba (II), and Cu (II)), which was produced via calcination [82]. Fu et al. developed a hybrid nanomaterial consisting titanate and lignin for the adsorption of Pb (II), Cu (II), and Cd (II) [120]. The hybrid nanomaterial exhibited an improved OH functional group when compared to singular nanomaterial, while exhibiting impressive reusability for HM adsorption. On the other hand, Yarandpour et al. developed a unique mesoporous poly (acrylic acid) (PAA)/dextran-polyaniline (PANI) core-shell nanofiber via the spinning process [121]. The hybrid nanofibers exhibited a fibrous morphology with a flake-like structure, developing a highly porous network throughout the nanomaterial. Subsequently, the nanofibers were able to adsorb more than 95% of Pb (II) ions from aqueous solution. Another hybrid material consisting of zeolite and silica oxide (SiO_2) was developed via wet chemical synthesis for the removal or HM in low concentrations [122]. The study suggests that both SiO_2 and zeolite work in tandem, where the outer layer consisting of SiO_2 rapidly adsorbs Zn (II) and Pb (II), and it then transports it via diffusion to its zeolite core, greatly enhancing the adsorption capacity. The employment of a non-metallic material that can produce a porous network was also the theme of the study done by Lui et al., where a biomimetic SiO_2@chitosan composite was effectively able to adsorb As (V) and Hg (II) when compared to singular nanomaterial [123].

4.3. Synthesis and Modification of Nanomaterials

The synthesis technique that was chosen for the formation of nanomaterials plays a significant role in the final characteristics exhibited. Generally, the synthesis route of nanomaterials follows two ways, which are top-down or bottom-up [110]. Figure 6 illustrates the building mechanism of the nanomaterial on both routes.

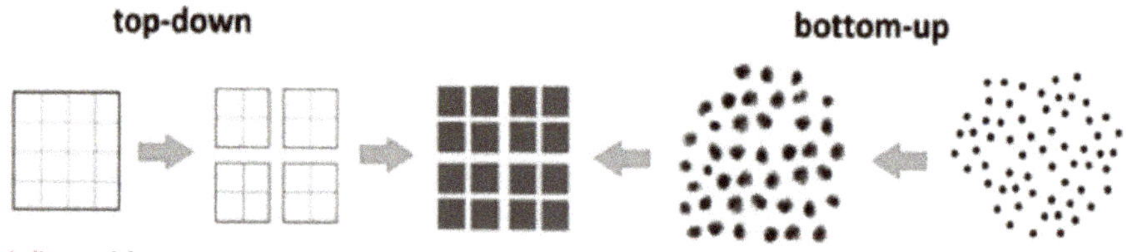

Figure 6. Nanomaterial synthesis route of nanomaterials following top-down, or bottom-up. Adapted from [107], with permission from CHEMIK, 2014.

The synthesis technique controls the design and building mechanism of nanoparticles to produce materials of precise structure, shape, and size. Some of the synthesis technique includes wet chemical synthesis, the solution combustion technique, hydrothermal technique, sol-gel technique, and mechanical milling [110]. However, there are three prominent methods i.e., sol-gel, hydrothermal, and wet chemical synthesis. These methods are prominent due to the fact that they are easily produced and facile, do not require expensive equipment's, and are highly reproducible. Sol-gel is one of the cheapest ways to produce nanomaterials with precise control on the stoichiometry of nanomaterial. The nanomaterials that were synthesized in this method generally had a small particle size and narrow size distribution due to their linear growth rate across the gel. As the name suggests, the precursors, commonly being metal alkoxides or metal chlorides, are hydrolyzed with water and alcohol and are then mixed and allowed to form a gel-like structure, before proceeding with calcination or sintering to remove the gel structure, leaving defined nanomaterials. Luu et al. synthesized a hybrid material that consisted of iron and TiO_2 via the sol-gel technique. The formed nanoparticle exhibited a size of 19.5 nm and a surface area of 42.9 m^2/g [124]. Another study for the formation of the same nanomaterial that was produced by Luu et al. exhibited a particle size of in the range of 19 to 7 nm and a surface area of between 84 and 182 m^2/g. The results indicated that the particle size highly governed the surface area, where smaller nanoparticles possessed a larger surface area [125]. Aware et al. produced a hybrid material of Zn and TiO_2 via the sol-gel method. The nanoparticle that was produced exhibited a particle size in the range of 12.6 to 18.1nm and a surface area value between 43.376 and 63.667 m_2/g [126]. Another popular method in nanomaterial synthesis is a hydrothermal technique. This is a versatile and simple method in the synthesis of nanomaterial under the high pressure and temperature closed condition. Precursors are commonly stirred with a strong alkali solution (NaOH) and are stirred until the precursors have reacted with the alkali reagent. The materials are then transferred into a Teflon lined autoclave stainless steel block and reacted under a predetermined temperature and time. The produced sample will then be washed with water to remove the excess sodium hydroxide (NaOH), and then dried and ground to obtain a high yield of nanomaterials. This method is popular because of its high degree of controllability [127]. Nanomaterials of different physical properties, including size, structure, and surface area can be easily obtained by small tweaks in the synthesis parameters. TiO_2, which has a globular shape, can be changed into TiO_2 nanotubes via the hydrothermal method, increasing its surface area from 40 m^2/g to 200 m^2/g [128]. Hydrothermal can also be used to produce hybrid nanomaterials in a facile way. The versatility of the hydrothermal technique allows it to form metal and non-metallic hybrid nanomaterials. Zhang et al. produced a hybrid nanomaterial that consists of Ag and nanocellulose, which exhibited high bactericidal efficiency against both bacteria and fungus with an average nanoparticle size of 86 nm [129]. Hydrothermal can also provide an alternative for synthesis of novel material, such as graphene oxide (GO), which is commonly produced via the Hummer's method [130]. In line with the simplicity of the synthesis method, wet chemical synthesis, or also known as the liquid phase synthesis technique, has also garnered the interest of researchers due to its simplicity while providing a large degree of control on the final physical characteristics of nanomaterials produced [131]. The process commonly involves the mixture of solutions of different ions at quantified volumes or molarity and provided a controlled amount of heat to initiate the formation of nanomaterials via precipitation. The excess solution is then washed away, and the precipitated nanomaterial is dried and ground/milled. This method offers control in nanomaterial stoichiometry, similar to the sol-gel method, and it can be conducted at a low temperature, reducing the energy consumption [131]. Table 3 shows some of the recent literature on the synthesis of hybrid nanomaterials using various techniques.

Table 3. Synthesis of hybrid nanomaterials using different techniques.

Product	Materials	Method	Parameters	Nanomaterial Characteristics	Reference
Mn-g-C_3N_4	Mn and g-C_3N_4	Combustion	Heated at a rate of 520 °C with a rate of 4 °C/min (2 h)	Improved ROS creation, lower band gap (1.25 eV)	[132]
Ag-GO	Ag and reduced GO	Modified Tour's method	Oxidized under 15 °C, heated to 50 °C, washed and freeze-dried	Visible light absorption improved oxidant generation capacity	[133]
CNT@MoS^2/SnS_2 nanotubes	CNT, MoS_2, and SnS_2	Hydrothermal	Autoclave for 180 °C for 20 h washed with water	Faster reduction of Cr (VI), the narrow bandgap	[134]
SnO_2-SrO	SnO_2 and SrO	Sol-gel	Gel formed, digested and dried in an oven at 100 °C, washed with ammoniated water	Lower band gap (2.23 eV, impart gas sensing	[135]
BiOBr/Ti_3C_2	BiOBr Ti_3C_2	Self-assembly method	BiOBr and Ti_3C_2 co-precipitated under magnetic stirring	Visible light photodegradation, surface reactivity	[98]
Yttrium/H-titanate	Yttrium and TiO_2	Hydrothermal	$Ti(SO_4)_2$ and hydrazine hydrate ($N_2H_4 \cdot H_2O$) reacted in an autoclave for 130 °C	Reduction in the efficiency of charge separation	[136]
TiO_2-MgO	(titanium isopropoxide and magnesium methoxide	Sol-gel	Sol-gel formed, dried at 100 °C and calcined at 900 °C	Visible light sensitivity, uniform hybrid material	[137]

5. Recent Progress and Performance Evaluation

5.1. Adsorption of Heavy Metal

Research that was conducted by Mousavi et al. was successful in developing a novel bifunctional ordered mesoporous silica via wet chemical synthesis, creating a porous nanomaterial with a highly arranged network of compact silicone group (S-O) [138]. The developed nanomaterial was able to adsorb 98.6% of Cu (II) and 98.09% of Zn (II) ions. Sharma et al. successfully synthesized a hybrid ZnO and TiO_2 monolith via the nanocasting technique. The hybrid material was able to adsorb Cd (II) via monolayer adsorption with an adsorption capacity of 786 mg/L, as compared to 643 mg/L for ZnO only monolith [83]. Castro et al. produced a biogenic iron compound via HM adsorption using biogenic iron sourced from mining wastewater that was able to absorb As (97.9 mg/L), Cr (20.1 mg/L), Zn (60.3 mg/L) and Cu (95.5 mg/L) in the optimum condition [84]. Activated carbons that were produced from nutshell removed almost 100% of Pb (II), 90–95% of Cu (II), and 80–90% of Zn (II), while biochar that was produced by pyrolysis of waste sludge was able to remove Cu (II), Cd (II), and Pb (II) by more than 90%. Based on the current research trends, researchers are exploring new and environmentally friendly carbon source to produce nano-sized carbon-based adsorbents that can improve HM adsorption due to the key features, such as the high surface area and porosity, ease for surface alkali functionalization, and being economically and environmentally viable. Deng et al. prepared a review on the utilisation of waste, such as palm shell, waste plastic, biomass, and so on to develop carbon based nanomaterials, including CNT and graphene [139]. Another emerging trend among research in the development of efficient adsorbents are the fabrication of various hybrid materials, which is expected to provide a synergistic effect in enhancing particle HM adsorption capacity. Kumar et al. developed hybrid carbon nanofibers and TiO_2 polyacrylonitrile (PAN) membranes via electrospinning to adsorb HM ions, such as Pb (II), Cu (II), and Cd (II) [140]. The developed CNFs/TiO_2–PAN hybrid membranes exhibited maximum adsorption of around 87%, 73%, 66%, for Pb (II), Cu (II), and Cd (II) metal ions, respectively. Another hybrid material comprised of Fe_3O_4 and *Raphiafarinifera*, a mangrove plant developed via the chemical co-precipitation method exhibited superior adsorption for Pb (II), Cu (II), Ni (II), Zn (II), and Cd (II) [141]. The hybrid material showed greater HM ion adsorption when compared to the singular particles, exhibiting the synergistic effect of the hybrid nanomaterial. Jiang et a. developed a hybrid graphitic carbon nitride nanosheet for the adsorption of both cationic and anionic HMs from wastewater [89]. The maximum adsorption capacities of Cd (II), Pb (II), and Cr (VI) on the g-C_3N_4 nanosheets are 123.205 mg/g, 136.571 mg/g, and 684.451 mg/g, respectively. The developed adsorbent can also be reused up to 10 times while achieving an adsorption capacity of 80% or more in 10 cycles.

Table 4 tabulates the synthesis technique and features that are imparted due to the formation of nano-sized adsorbents, which includes improved electrical conductivity, partial cation exchange, highly selective adsorption, high particle porosity, and surface chemical activation. An increase in surface area allows for more adsorption sites, which directly increases adsorption capacity. Porous structure also exhibits the same feature, in which the availability of binding sites for HM ions increases with the increasing porous structures. Nanoparticles are also functionalized to produce highly selective adsorbents, which are capable of adsorbing particular HM ions in a mixed metal water system. Chen et al. demonstrated that, by doping SiO_2 with polythiophene (PTh), the composite recorded an impressive selectivity towards Zn (II) ions in multiple ion solutions (with p-value > 0.8 as compared to 0.3 of Pb (II)), which has also increased the adsorption capacity [142]. It was postulated that PTh has an affinity towards Zn (II) ions, which contributed to the overall selectivity. These features are important in improving the adoption of cationic HMs, such as increasing the strength of adsorption and increasing the adsorption capacity due to the porous structure of nanoparticles as adsorbents. Additionally, literature showed that there are prominent metal ions that are consistently used as model ions, including Cu (II), Pb (II), Co (II), Cr (II), and Zn (II) [143]. Other notable HM ions that are studied upon include Li (I), As (II), and Ni (II). In regards to adsorption capacity, the results vary, which is expected when presuming that all of the nanomaterials produced have varying features that highly influence the adsorption of specific HMs. The employment of hierarchically mesoporous carbon was able to adsorb Cu (II) up to 215.0 mg/g, while PTh/SiO_2 nanoparticles were only able to adsorb 35.3 mg/g of Cu (II) [142].

A hybrid nanoadsorbent consisting of Fe and GO was synthesized to produce a magnetic graphene oxide (MGO) [37]. The MGO exhibited superior adsorption of Cd (II) and As (V) as compared to singular GO. The superiority of MGO was mainly attributed to its high dispersibility, thin nanosheets that were exhibited by GO, the synergistic effect that resulted from the electrostatic attraction offered by Fe, and various O-containing functional groups due to the surface functionality of GO. Marciniak et al. synthesized oxidized mesoporous carbon nanoparticle via the hart template method for the adsorption of Ni (II) and Cd (II) [144]. The nanoparticles were oxidized in different degrees to vary its oxidizing functional group. The results indicated that the nanoparticle with highest surface functional groups of acidic character was able to adsorb more HM ions. Similarly, research that was conducted by Li et al., where they prepared a sludge-based activated carbon impregnated with HNO_3 for the removal of Pb (II), was highly governed by the surface functional group [145]. The results indicated that modified adsorbent rich with carboxyl group (R-COO^-) was able to adsorb up to 98% of Pb (II), as compared to an unmodified particle which was only able to adsorb 83% of Pb (II). Huang et al. prepared a core-shell Fe_3O_4@polytetramethylene terephthalate (PTMT) composite magnetic microspheres for the adsorption of HM in highly saline water [38]. Surface amine functionalization greatly reduced particle agglomeration, which was complicated by nanomaterials with large surface area, while the incorporation of magnetite enhanced the recovery efficiency of the adsorbent while using a simple magnet. In addition, the shell structure also increased the particle surface area and porosity, increasing the adsorption sites, which improved the adsorption capacity of Hg (II) and Pb (II). Evidence of the impact that unique structure has on the adsorption of HM can also be seen in the study done by Ma et al., where a waste cotton fabric based double network hydrogel was developed for HM removal [146]. The experimental data suggested that the porous and sheet-like laminar structures that were exhibited by the nanostructured adsorbent were the reason behind the fast kinetics of Cu (II) and Cd (II) sorption equilibrium displayed. Similarly, Wang et al. developed a hybrid graphene oxide/silk fibroin hybrid aerogel [147]. The synthesized hybrid nanomaterial exhibited a porous network, which helped the HM ions to easily diffuse into the aerogels, while the silk fibroin exhibited great chelating feature, holding onto the HM ions for excellent adsorption capacity (Ag (II): 195.8 mg/g, Cu (II): 72.1 mg/g, Cu (II): 83.4 mg/g). Table 4 shows a compilation of recent literature on the development of various types of adsorbent for the removal of various HM ions from wastewaters.

Table 4. Recent literature on the development of various types of adsorbent for the removal of HM ions from wastewaters.

Nanomaterial	Synthesis Technique	Features	Metal Species	Adsorption Capacity	Optimum pH	Reference
Mesoporous carbon	Hard template technique	High uniformity of porous structure, surface functionalized	Ni (II) Co (II)	140.9 mg/g 129.9 mg/g	5	[144]
Hierarchically porous carbon	Pyrolysis and chemical activation	KOH activated	Cd (II) Pb (II) Cu (II) Zn (II) Cr(III)	180.0 mg/g 220.0 mg/g 215.0 mg/g 95.0 mg/g 140.0 mg/g	6	[40]
Geopolymers	Alkali activation of aluminosilicate	Alkali activated, inorganic polymers	Ni (II) Pb (II) Cd (II)	85.3 mg/g 111.0 mg/g 130.5 mg/g	10	[148]
Polyaniline/TiO_2	Chemical oxidative polymerisation	Self-doping, highly selective adsorption	Zn (II) Pb (II) Cu (II)	51.6 mg/g 96.2 mg/g 18.2 mg/g	5	[45]
Ga-doped ZnO	Sol-gel	Improved electrical conductivity	Cr (II) Cd (II)	52.2 mg/g 28.3 mg/g	3–5	[149]
Polythiophene/SiO_2	Sol-gel	Stable and highly selective	Pb (II) Cu (II) Zn (II)	70.9 mg/g 35.3 mg/g 34.6 mg/g	5	[142]
Zeolite	Amino acid as mesoporogens	Partial cation exchange	Cu (II) Ni (II) Pb (II)	171 mg/g 99.1 mg/g 514.0 mg/g	11	[46]
Wild herb nanoparticle	Ball milling	Environmentally friendly	Cd (II) Co (II) Li (II)	52.9 mg/g 40.8 mg/g 181.8 mg/g	12	[39]
Fe_3O_4	Carbon microsphere	High BET surface area, hierarchical as well as mesoporous structures	Pb (II) Cd (II) Cr (III)	95.2% 96.2% 98.2%	6	[150]

5.2. Photocatalysis of Heavy Metal

Recently, an iron oxide (II) bismuth carbonate hybrid photocatalyst was developed and it showed excellent photocatalytic activity towards the reduction of carcinogenic and mutagenic Cr (VI) to nontoxic Cr (II) [151]. Another study that was done by Du et al. showed excellent photocatalytic Cr (VI) reduction to Cr (III) [152]. Kumar et al. employed a hybrid WO_3/reduced graphene oxide (rGO) nanocomposites photocatalyst that exhibited a photocatalytic reduction of Cr (VI) following first-order kinetics and rate constants were found to be 0.0084 min^{-1} [153]. Even though traditional semiconducting photocatalyst exhibited good HM ion reduction, they are exclusively responsive towards dedicated UV light sources due to their large band gap. In order to overcome this, traditional photocatalyst were doped/modified with metallic or non-metallic materials to impart visible light sensitivity, producing nanocomposite photocatalyst. The presence of a froing atom on the surface photocatalyst also reduces the band gap, which reduces the energy that is required to excite an electron for photocatalyst initiation. It also increases the rate of photocatalysis due to the presence of a charge carrier [154]. This allows the photocatalyst to absorb photons from visible light sources to excite an electron and initiate the photocatalytic reaction. Figure 7 shows the photoreduction performance of Cr (VI) in a single system and in coexistence with an organic pollutant.

Lie et al. conducted a study on the simultaneous photoreduction of Cr (VI) and the photodegradation of Rhodamine B (RhB) to in an independent and dependent system. The results that are depicted in Figure 7a,b indicates that the photoreduction of Cr (VI) in the presence of RhB is significantly better when compared to photoreduction in a solution only containing Cr (VI) ions [155]. The research work for the simultaneous degradation of phenol and photoreduction of Cr (VI) that was conducted by Yu et al. explained that the presence of phenols improves the photoreduction of Cr (VI) as the electron-hole is consumed by phenol. This suppresses the recombination of electron and holes, which in turn accelerates the photoreduction of Cr (VI) [48]. The same reasoning has been used in other research, where the photoreduction of Cr (VI) was accompanied by the degradation of various types of organic pollutants to sustain the reaction, such as 2-Mercaptobenzothiazole (MBT) [156], sulfamethoxazole [157], RhB [98,136], and chlorinated phenols [158]. The photoreduction of both Pb (II) and Cu (II) was attempted together with the photodegradation of RhB while using a modified Chitosan-Gelatin @ zirconium (IV) selenophosphate nanocomposite ion exchanger. The hybrid photocatalyst was able to retain more than 90% of the HMs, while degrading more than 80% of RhB within 120 min [159]. Another example was exhibited in a research work that was conducted by Du et al. Their investigation showed that the addition of different organic compounds, like citric acid, oxalic acid, and diclofenac sodium mimics the function of a hole scavenger, which increases the photocatalytic Cr (VI) reduction activity. The ability of hole scavengers to consume the photoinduced holes that are produced by the photocatalyst upon light irradiation is accepted as a plausible reason. This allows for more electrons to escape from pair recombination and become available for the reduction of Cr (VI) ions [152]. Photocatalysis have also shown the potential to photoreduce Cr (VI) into Cr (III) while exhibiting potential to kill harmful bacteria, such as *E. coli* (89%) and *S. aureus* (81%) within 10 min of irradiation on photosensitized TiO_2 nanofibers [160]. In recent years, the development of binary (bi-material) and ternary (tri-material) photocatalyst has intensified in search of combinations to produce photocatalyst that is responsive towards visible light, low recombination rates, and increased surface reactive sites.

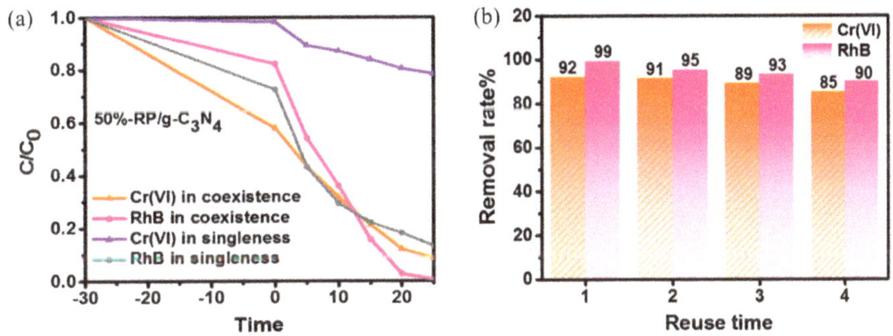

Figure 7. (**a**) Photoreduction of Cr (VI) in the presence and absence of Rhodamine B (RhB) and (**b**) removal rate of both Cr (VI) and RhB at different individual cycles. Adapted from [155], with permission from Elsevier, 2019.

An aggregation free CeO_2/SnO_2/rGO was developed via the hydrothermal process for the simultaneous degradation of MB and photoreduction of Pb (II) and Cd (II). The photocatalyst exhibited close to 100% degradation of methylene blue (MB) and a HM ion reduction of close to 80%. The rGo acts as a charge suppressor and, due to its excellent electrical conductivity, it can trap the electron that is excited from the conduction band (CB) of SnO_2 and CeO_2, allowing for longer photocatalyst activation. The exploitation of rGO, as an excellent charge separator and trapping site to sustain photocatalytic formation of ROS, was also explored by Bai et al., where a Red Phosporus (RP)-MoS_2/rGO ternary photocatalyst was developed via a facile two-step hydrothermal [161]. RP and MoS_2 exhibited excellent photocatalytic activity and increased the number of excited electrons/holes under visible light irradiation, while rGO was responsible for enhancing charge separation to sustain and provide stability in electron recombination. These factors synergistically assisted RP-MoS_2/rGO to perform impeccably, reducing Cr (VI) by up to 95%. Another nanomaterial that exhibits charge trapping capability is carbon nitride (C_3N_4), which is a relatively new nanomaterial. Liu et al. developed a binary RP/g-C_3N_4 photocatalyst for visible light photoreduction of Cr (VI). The catalyst exhibited good stability for removing 85% Cr (VI) and 90% RhB, even after four times of recycling, owing to the effective separation and rapid transfer of e^-/h^+ pairs between RP and g-C_3N_4. With the low band gap exhibited (1.35 eV), $\bullet O^{2-}$ can be generated in this system for the lower CB position of RP, where the excited electron can jump into the CB of RP, sustaining photocatalytic activity for a longer period. The synergistic redox reaction allows for Cr (VI) to reduce into Cr (III) [155]. Ye et al. managed to synthesize a ternary $Ag/Bi_4O_7/g\text{-}C_3N_4$ nanosheets Z-scheme heterojunction photocatalyst for the photoreduction of Cr (VI) while using a combination of thermal polymerization, hydrothermal, and calcination [162]. The structure exhibited the deposition of Ag and Bi_4O_7 on sheets of $g\text{-}C_3N_4$, similarly to the work that was conducted by Liu et al. The relatively low band gap of both Bi_4O_7 (1.89 eV) and $g\text{-}C_3N_4$ (2.97 eV) enabled the ternary photocatalyst to exhibit excellent visible light responsive capabilities. Additionally, the presence of $g\text{-}C_3N_4$ sheets offered excellent charge separation capabilities, thus enhancing the photoreduction of Cr (VI) [163]. Jing et al. developed a three-dimensional $PANI/MgIn_2S_4$ nanoflower photocatalyst via electrostatic adsorption between PANI and $MgIn_2S_4$. PANI is known to be a good polymeric electric conductor due to the presence of heteroconjugated π bond, making it a suitable candidate as a dopant for photocatalysts [164]. The ternary photocatalyst was able to reduce Cr (VI) by up to 95% within 15 min of light irradiation. The photogenerated electrons in the valence band of $MgIn_2S_4$ and the highest occupied molecular orbital (HOMO) of PANI could be excited to the corresponding conduction band and the lowest unoccupied molecular orbital (LUMO) when it assimilates photons with energy larger than the band gap, while the holes were generated in the valence band of $MgIn_2S_4$ and the HOMO of PANI. This interaction accelerated the photoreduction of Cr (VI) and the overall photoactivity [165]. Table 5 shows the development of nanocomposite photocatalyst for the reduction of various types of HM.

Table 5. Development of nanocomposite photocatalyst for the reduction of various types of HM.

Photocatalyst	Dopant	Method	Metal Species	Removal Performance	Reference
$Ag/Bi_4O_7/$	$g\text{-}C_3N_4$	Thermal polymerization, hydrothermal and calcination	Cr (VI)	90%	[162]
WO_3	Reduced graphene oxide (rGO)	In-situ hydrothermal	Cr (VI)	90%	[153]
Fe_2O_3	Bismuth carbonate (BOC)	Two-step chemical modification	Cr (VI)	>90%	[151]
TiO_2	Graphene	Hydrothermal	Pb (II)	60%	[166]
V_2O_5 nanorod	$g\text{-}C_3N_4$ nanosheets	Facile impregnation	Cr (VI)	71%	[167]
Zirconium	Selenophosphate	Two-step ion exchanger	Pb (II) Mg (II)	100% 95%	[159]
Red phosphorus	$g\text{-}C_3N_4$ nanosheets	Thermal polymerization and hydrothermal	Cr (VI)	92%	[155]
TiO_2	-	-	Cd (II) Pb (II)	98% 99%	[168]
Metal organic framework 100	g-C3N4 nanosheets	Calcination and hydrothermal	Cr (VI)	98%	[152]
Zn	Coordination polymers (H_2L and by)	Hydrothermal	Cr (VI)	100%	[169]
CdS	CuInS	Hydrothermal	Cr (VI)	100%	[170]
Titanate nanosheets	Yttrium	Hydrothermal	Cr (VI)	>75%	[136]
TiO_2	Graphene	Hydrothermal	Zn (II)	100%	[171]
TiO_2	Graphene	Hydrothermal	Pb (II)	>70%	[166]
$CeO_2/SnO_2/$	rGO	Hydrothermal	Pb (II) Cd (II)	80% 80%	[172]

5.3. *Membrane Composite for Removal of Heavy Metal*

Nanomaterials are commonly added into polymeric membranes during preparation, where it is dispersed into the dope solution before the membranes are formed via the dry-wet phase inversion technique. The dispersion of nanomaterials across the membrane matrix or deposition on the membrane surface, as in the case of thin-film nanocomposite (TFN) membrane, has significantly altered the permeation and rejection performance of the membranes. Lakhotia et al. developed a TFN membrane with FeO nanoparticle dispersed on the membrane selective layer. As the concentration of FeO nanoparticles increased on the membrane surface, the hydrophilicity and surface charge of the nanocomposite membranes were effectively enhanced. The membrane flux was enhanced from 27.46 to 36.85 L/m^2 h and high rejection of Mg (II) and Na (I) rejection (>90%), whilst membrane without the FeO nanoparticle rejected salt in the range of 65%. An $AlTi_2O_6$ incorporated polysulfone (PSF) composite membrane exhibited improved hydrophilicity with the water contact angle being reduced from 73° to 51° [173]. The presence of $AlTi_2O_6$ hindered the flow of non-solvent during membrane casting, which created more membrane pores on the surface. Polypyrrole (PPy)@Al_2O_3 was added into a polyethersulfone (PES) membrane matrix for the removal of Cu (II). The addition of PPy@Al_2O_3 improved the membrane water transport capacity, reduced membrane surface roughness, and eventually mitigated membrane fouling [174]. The nanocomposite membrane also exhibited improved Cu (II) rejection (25% to 81%). Ghaemi et al. developed a polyaniline modified GO nanoparticle that was incorporated into a PES membrane for the remediation of water laden with Pb (II), which also showed good metal ion removal because the incorporating of PANI modified GO increased the viscosity of the membrane dope solution, which in turn reduced the mean pore size of nanocomposite membranes [175]. The filtration of HM-laden water produced clear water containing trace levels of HM as permeate, in which the HMs were concentrated in the retentate.

A study done by Nasir et al. showed that the incorporation of hydrous iron manganese nanoparticle into a PSF membrane was able to adsorb As (II) up to 4.1. mg/g [176]. A zirconia polyvinylidene fluoride (PVDF) composite membrane that was developed by Zheng et al. was able to adsorb As (V) at 25.5 mg/g, whilst maintaining membrane flux of 177.6 L/m^2 h [69]. Another work that was done by He et al. incorporated zirconia into a PSF hollow fiber membrane for the removal of As (V) [177]. This work exhibited a much higher adsorption capacity of 131.8 mg/g at zirconia/polymer ratio of 1.5. In a separate study, zirconia was modified while using phosphate solution to produce a PVDF membrane with surface coated nanomaterial that performed excellently for the removal of Pb (II) ions [178]. The adsorption capacity that was recorded by the optimized membrane was 121.2 mg/g at the pH of 5.5. The modification also allowed for selectively improving the membrane adsorption to favor Pb (II) over Zn (II) ions. Delavar developed HMO nanoparticle incorporated membrane for the removal of Cd (II) and Cu (II), where it was revealed that the loading of nanomaterial into the membrane matrix largely influences the adsorption performance. The polymeric structure of the membrane only serves as a pathway for untreated water to pass and to immobilize the adsorbents. Membrane that was loaded with 3 wt% adsorbed 31.3 mg/g of Cd (II) and 29.7 mg/g of Cu (II), whilst 10% loaded membrane adsorbed 33.18 mg/g and 30.6 mg/g of Cd (II) and Cu (II), respectively, asserting this notion [179]. In addition to this, another work on the employment of HMO as a nanoadsorbent, immobilized in a PES membrane, was able to adsorb Pb (II) at an uptake capacity of 204 mg/g, highlighting the potential of HMO as an excellent candidate for HM adsorption [180]. Table 6 shows the details of membranes incorporated with nanomaterials that were developed for the rejection of HMs.

Table 6. Membranes incorporated with nanomaterials for HM ion removal.

Polymer	Nanomaterial	Removal Method	Metal Species	Control Membrane Performance	Composite Membrane Performance	Reference
Polyethylene oxide (PEO)	Halloysite nanotubes and	Adsorption	Cr (VI) Cd (II) Cu (II) Pb (II)	80 mg/g 105 mg/g 120 mg/g 145 mg/g	85 mg/g 115 mg/g 135 mg/g 155 mg/g	[181]
Ceramic	Rice husk ash	Adsorption and Filtration	Ni (II) Zn (II) Pb (II)	-	99.99% 99.97% 99/99%	[182]
Polyvinyl chloride (PVC)	Carboxylated CNT	Filtration	Zn (II)	48%	93%	[183]
PVDF	Superhydrophilic alumina	Filtration	Pb (II)	84%	92.5%	[184]
Alumina Substrate	Zeolite imidazolate framework-30	Filtration	Cu (II)	-	99.87%	[185]
Polyethylenimine (PEI)	GO	Filtration	Zn (II)	0%	96.6%	[186]
Poly(ethyl methacrylate) PEMA	Rhodanine	Adsorption	Ag (II) Pb (II)	- -	65% 58%	[65]
PSF	Nickel/Iron oxide	Filtration	Pb (II) Cu (II)	- -	95% 95%	[65]
PAN	HMO	Filtration	Cu (II)	37%	70%	[187]
PES	Polydopamine	Adsorption	Pb (II) Cu (II) Cd (II)	>1 mg/g >1 mg/g >1 mg/g	20.3 mg/g 10.4 mg/g 17 mg/g	[56]
PSF	Al-Ti_2O_6	Filtration	As (II) Cd (II) Pb (II)	- - -	96% 98% 99%	[173]
PSF	Quaternized polyelectrolyte complex	Filtration	Mg (II) Zn (II) Cu (II)	86.4% 87.1% 80.3%	95.7% 98.3% 97.9%	[57]

6. Conclusions and Future Perspective

The advances of nanomaterial in terms of development of materials with novel structure, characteristics, and hybrid nanomaterials have further revealed unique and amplified properties. The ability to develop nanomaterials atom by atom with highly controllable methods is behind the rise of novel nanomaterials. The review highlights the roles of nanomaterials in heightening the efficiency of the above-mentioned technologies for HM removal. However, there is plenty of room for future development and research to discover new hybrid nanomaterials for the removal of HM from wastewaters. Removing HM ions from water sources is a critical task. Removing and destroying HM is difficult due to its stability in ionic form. However, recent advances in using various types of nanomaterials that were developed with HM removal indicate the potential that researchers see in nanomaterials. Of all the proven methods to remove HM, three prominent ways i.e., adsorption, photocatalysis, and membrane separation have been comprehensively reviewed. Undeniably, nanomaterials have enabled significant breakthrough that was made in these strategies. For the adsorption of HM, researchers have developed plenty of nanomaterials, singular or hybrid, which work as nanoadsorbent to trap HM from water bodies, as displayed in Table 4. Even though all of the nanomaterials excel as absorbents, more emphasis needs to be placed in the development of functional HM adsorbents using waste products via the facile synthesis method. The reason behind this notion is that greater research emphasis is required to produce nanomaterials that can be easily scaled up for real-life usage. For instance, producing a very large amount of absorbents for effective treatment of large water bodies, such as lakes and rivers. Current trends in developing adsorbents from waste, such as from biochar or sludge waste, can be studied in depth in terms of upscaling. It needs to be highlighted that, whilst it is great for researchers to pursue unique and novel nanomaterial with new features that can aid in HM removal, it is important to consider the effectiveness of the approaches for the large-scale removal of HM. Photocatalysis has shown great promise in reducing HM ions while destroying organic pollutants, which exhibits the versatility of semiconducting nanomaterial as the total remediation of wastewater. Great strides have been taken in sensitizing traditional semiconductors to respond towards visible light source by doping. However, doping needs to be focused on utilizing non-noble metals and non-metal dopants, which are much more abundant in nature when compared to rare and noble metals. In addition, future research to be conducted can place better emphasis on the doping method to improve the precision of dopant incorporated into semiconductor nanomaterials. On the other hand, the feasibility of employing plasmonic metal nanomaterial for the reduction of HM ions should also be explored. Plasmonic metal nanoparticles are known as light harvesting materials that exhibit the ability to harvest visible light photons through the excitation of localized surface plasmon resonance (LSPR). Currently, the studies on this nanomaterial to reduce HM ions in waterways are very scarce. The unique characteristic of the plasmonic metal nanoparticle is expected to spur interests in this field [188,189]. The incorporation of functional nanomaterials has enhanced polymeric membrane separation and permeation efficiency, where membrane separation or adsorption are two viable ways of removing HM ions from water sources. However, the agglomeration of nanomaterial in the membrane matrix is a common problem, which limits the loading of nanomaterial and reduces the efficiency of nanomaterials. To counter this, steps to functionalise the nanomaterial surface to inhibit agglomeration (employing silane agents, improved surface charge) is welcome progress in further maximizing the synergistic of nanomaterial and membrane technology.

Author Contributions: M.N.S. and P.S.G. conceived, designed and wrote the manuscript, while W.J.L. and A.F.I. were responsible to visualization, editing and funding acquisition.

Funding: This research was funded by the Ministry of Higher Education (HiCOE Grant, Vot no; 4J183, Fundamental Research Grant Scheme (FRGS, Vot no; 5F005).

Acknowledgments: The authors would like to express their sincere gratitude to Advanced Membrane Technology Research Center (AMTEC) and University Teknologi Malaysia for their support.

Conflicts of Interest: The authors declare no conflict of interest.

References

1. Carolin, C.F.; Kumar, P.S.; Saravanan, A.; Joshiba, G.J.; Naushad, M. Efficient techniques for the removal of toxic heavy metals from aquatic environment: A review. *J. Environ. Chem. Eng.* **2017**, *5*, 2782–2799. [CrossRef]
2. Duncan, A.E.; de Vries, N.; Nyarko, K.B. Assessment of heavy metal pollution in the sediments of the River Pra and its tributaries. *Water Air Soil Pollut.* **2018**, *229*, 272. [CrossRef]
3. Thakur, L.S. Heavy metal Cu, Ni and Zn: Toxicity, health hazards and their removal techniques by low cost adsorbents: A short overview. *Int. J. Plant Sci.* **2015**, *3*, 143–157.
4. Akpor, O.B.; Ohiobor, G.O.; Olaolu, T.D. Heavy metal pollutants in wastewater effluents: Sources, effects and remediation. *Adv. Biosci. Bioeng.* **2014**, *2*, 37–43. [CrossRef]
5. Yi, Y.; Tang, C.; Yi, T.; Yang, Z.; Zhang, S. Ecotoxicology and environmental safety health risk assessment of heavy metals in fi sh and accumulation patterns in food web in the upper Yangtze River, China. *Ecotoxicol. Environ. Saf.* **2017**, *145*, 295–302. [CrossRef] [PubMed]
6. Madzin, Z.; Kusin, F.M.; Yusof, F.M.; Muhammad, S.N. Assessment of water quality index and heavy metal contamination in active and abandoned iron ore mining sites in Pahang, Malaysia. *MATEC Web Conf.* **2017**, *103*, 05010. [CrossRef]
7. Ganesapillai, M.; Thanabalan, M.; Anjum, H.; Arunagiri, A.; Johari, K.; Gnanasundaram, N.; Regupathi, I. A review on adsorptive removal of oil pollutants (BTEX) from wastewater using carbon nanotubes. *J. Mol. Liq.* **2018**, *277*, 1005–1025.

8. Lee, X.J.; Hiew, B.Y.Z.; Lai, K.C.; Lee, L.Y.; Gan, S.; Thangalazhy-Gopakumar, S.; Rigby, S. Review on graphene and its derivatives: Synthesis methods and potential industrial implementation. *J. Taiwan Inst. Chem. Eng.* **2018**. [CrossRef]
9. Ou, H.H.; Lo, S.L. Review of titania nanotubes synthesized via the hydrothermal treatment: Fabrication, modification, and application. *Sep. Purif. Technol.* **2007**, *58*, 179–191. [CrossRef]
10. Sheet, I.; Kabbani, A.; Holail, H. Removal of heavy metals using nanostructured graphite oxide, silica nanoparticles and silica/graphite oxide composite. *Energy Procedia* **2014**, *50*, 130–138. [CrossRef]
11. Masteri-Farahani, M.; Ghahremani, M. Surface functionalization of graphene oxide and graphene oxide-magnetite nanocomposite with molybdenum-bidentate Schiff base complex. *J. Phys. Chem. Solids* **2019**, *130*, 6–12. [CrossRef]
12. Khin, M.M.; Nair, A.S.; Babu, V.J.; Murugan, R.; Ramakrishna, S. A review on nanomaterials for environmental remediation. *Energy Environ. Sci.* **2012**, *5*, 8075–8109. [CrossRef]
13. Azzouz, A.; Kailasa, S.K.; Lee, S.S.; Rascón, A.J.; Ballesteros, E.; Zhang, M.; Kim, K.H. Review of nanomaterials as sorbents in solid-phase extraction for environmental samples. *TrAC Trends Anal. Chem.* **2018**, *108*, 347–369. [CrossRef]
14. Jeevanandam, J.; Barhoum, A.; Chan, Y.S.; Dufresne, A.; Danquah, M.K. Review on nanoparticles and nanostructured materials: History, sources, toxicity and regulations. *Beilstein J. Nanotechnol.* **2018**, *9*, 1050–1074. [CrossRef] [PubMed]
15. Santhosh, C.; Velmurugan, V.; Jacob, G.; Jeong, S.K.; Grace, A.N.; Bhatnagar, A. Role of nanomaterials in water treatment applications: A review. *Chem. Eng. J.* **2016**, *306*, 1116–1137. [CrossRef]
16. Ozuni, E.; Dhaskali, L.; Abeshi, J.; Zogaj, M.; Haz Iri, I.; Beqiraj, D.; Latifi, F. Heavy metals in fish for public consumption and consumer protection. *Nat. Montenegrina* **2010**, *9*, 843–851.
17. Krieger, R. *Handbook of Pesticide Toxicology*; Academic Press: Cambridge, MA, USA, 2010; Volume 3.
18. Maouni, A.; Lamarti, A.; Aidoun, A.; Khaddor, M.; Badoc, A. Effect of benzimidazole fungicides and calcium chloride on *Alternaria alternata* and *Penicillium expansum* rot during storage of pears. *Afr. J. Biotechnol.* **2007**, *6*, 1289–1292.
19. Viehweger, K. How plants cope with heavy metals. *Bot. Stud.* **2014**, *55*, 35. [CrossRef]
20. Popova, E.; System, S.; Popova, E. Accumulation of heavy metals in the "soil-plant" system. *AIP Conf. Proc.* **2016**, *1772*, 050006.
21. Nouri, J.; Khorasani, N.; Lorenstani, B.; Karami, M.; Hassani, A.H.; Yousefi, N. Accumulation of heavy metals in soil and uptake by plant species with phytoremediation potential. *Environ. Earth Sci.* **2009**, *59*, 315–323. [CrossRef]
22. Yozukmaz, A.; Sel, F. Heavy metal accumulation in the leaves, stem and root of the invasive submerged macrophyte *Myriophyllum spicatum* L. (Haloragaceae): An Example of Kadın Creek (Mugla, Turkey). *Braz. Arch. Biol. Technol.* **2014**, *57*, 434–440.
23. Vongdala, N.; Tran, H.; Xuan, T.D.; Teschke, R. Heavy metal accumulation in water soil, and plants of municipal solid waste landfill in Vientiane, Laos. *Int. J. Environ. Res. Public Health* **2019**, *16*, 22. [CrossRef] [PubMed]
24. Tangahu, B.V.; Rozaimah, S.; Abdullah, S.; Basri, H.; Idris, M.; Anuar, N.; Mukhlisin, M. A review on heavy metals (As, Pb, and Hg) uptake by plants through phytoremediation. *Int. J. Chem. Eng.* **2011**, *2011*, 939161. [CrossRef]
25. Gall, J.E.; Boyd, R.S.; Rajakaruna, N. Transfer of heavy metals through terrestrial food webs: A review. *Environ. Monit. Assesment* **2015**, *187*, 200–221. [CrossRef] [PubMed]
26. Suvarapu, L.N.; Baek, S.O. Determination of heavy metals in the ambient atmosphere: A review. *Toxicol. Ind. Health* **2016**, *33*, 79–96. [CrossRef]
27. Gimeno-García, E.; Andreu, V.; Boluda, R. Heavy metals incidence in the application of inorganic fertilizers and pesticides to rice farming soils. *Environ. Pollut.* **1996**, *92*, 19–25. [CrossRef]
28. Grinang, J.; Sim, S.F.; Rajendran, M.; Nyanti, L.; Ling, T.Y.; Liew, J.J. Assessment of trace metals in water and sediment in a tropical river potentially affected by land use activities in northern Sarawak, Malaysia. *Int. J. Environ. Res.* **2017**, *11*, 99–110.
29. Singh, N.; Gupta, V.K.; Kumar, A.; Sharma, B. Synergistic effects of heavy metals and pesticides in living systems. *Front. Chem.* **2017**, *5*, 1–9. [CrossRef]

30. Jaishankar, M.; Tseten, T.; Anbalagan, N.; Mathew, B.B.; Beeregowda, K.N. Toxicity, mechanism and health effects of some heavy metals. *Interdiscip. Toxicol.* **2014**, *7*, 60–72. [CrossRef]
31. Mathew, B.B.; Tiwari, A.; Jatawa, S.K. A review on free radicals and antioxidants. *J. Pharm. Res.* **2011**, *4*, 4340–4343.
32. Flora, S.J.S.; Mittal, M.; Mehta, A. Heavy metal induced oxidative stress & its possible reversal by chelation therapy. *Indian J. Med. Res.* **2008**, *128*, 501–523.
33. Ojezele, O.J.; Ojezele, M.O.; Adeosun, A.M. Cooking utensils as probable source of heavy metal toxicity. *Middle-East J. Sci. Res.* **2016**, *24*, 2216–2220.
34. Krewski, D.; Yokel, R.A.; Nieboer, E.; Borchelt, D.; Cohen, J.; Harry, J.; Kacew, S.; Lindsay, J.; Mahfouz, A.M.; Rondeau, V. Human Health Risk Assessment for Aluminium, Aluminium Oxide, and Aluminium Hydroxide. *J. Toxicol. Environ. Health B. Crit. Rev.* **2008**, *11*, 1–269. [CrossRef]
35. Bernhoft, R.A. Mercury toxicity and treatment: A review of the literature. *J. Environ. Public Health* **2012**, *2012*. [CrossRef] [PubMed]
36. Bhatt, P.; Rani, A. Textile dyeing and printing industry: An environmental hazard. *Asian Dye* **2013**, *10*, 51–54.
37. Huang, D.; Wu, J.; Wang, L.; Liu, X.; Meng, J.; Tang, X.; Tang, C.; Xu, J. Novel insight into adsorption and co-adsorption of heavy metal ions and an organic pollutant by magnetic graphene nanomaterials in water. *Chem. Eng. J.* **2019**, *358*, 1399–1409. [CrossRef]
38. Bi, J.; Huang, X.; Xie, C.; Yang, J.; Wang, J.; Hao, H. Design and synthesis of core–shell Fe_3O_4@PTMT composite magnetic microspheres for adsorption of heavy metals from high salinity wastewater. *Chemosphere* **2018**, *206*, 513–521.
39. Al-Senani, G.M.; Al-Fawzan, F.F. Adsorption study of heavy metal ions from aqueous solution by nanoparticle of wild herbs. *Egypt. J. Aquat. Res.* **2018**, *44*, 187–194. [CrossRef]
40. Sun, J.; Li, M.; Zhang, Z.; Guo, J. Unravelling the adsorption disparity mechanism of heavy-metal ions on the biomass-derived hierarchically porous carbon. *Appl. Surf. Sci.* **2019**, *471*, 615–620. [CrossRef]
41. Sing, K.S.W.; Everett, D.H.; Haul, R.A.W.; Moscou, L.; Pieroti, R.A.; Rouquerol, J.; Siemieniewska, T. Reporting physisorption data for gas/solid systems with special reference to the determination of surface area and porosity. *Pure Appl. Chem.* **1985**, *57*, 603–619. [CrossRef]
42. Berger, A.H.; Bhown, A.S. Comparing physisorption and chemisorption solid sorbents for use separating CO_2 from flue gas using temperature swing adsorption. *Energy Procedia* **2011**, *4*, 562–567. [CrossRef]
43. Kim, W.-H.; Chung, H.-K.; Park, J.; Park, P.-K.; Cho, J.; Jeong, T.-Y. Application of Langmuir and Freundlich isotherms to predict adsorbate removal efficiency or required amount of adsorbent. *J. Ind. Eng. Chem.* **2015**, *28*, 241–246.
44. Ghosal, P.S.; Gupta, A.K. Development of a generalized adsorption isotherm model at solid-liquid interface: A novel approach. *J. Mol. Liq.* **2017**, *240*, 21–24. [CrossRef]
45. Chen, J.; Wang, N.; Liu, Y.; Zhu, J.; Feng, J.; Yan, W. Synergetic effect in a self-doping polyaniline/TiO_2 composite for selective adsorption of heavy metal ions. *Synth. Met.* **2018**, *245*, 32–41. [CrossRef]
46. Hong, M.; Yu, L.; Wang, Y.; Zhang, J.; Chen, Z.; Dong, L.; Zan, Q.; Li, R. Heavy metal adsorption with zeolites: The role of hierarchical pore architecture. *Chem. Eng. J.* **2019**, *359*, 363–372. [CrossRef]
47. Ni, B.; Huang, Q.; Wang, C.; Ni, T.; Sun, J.; Wei, W. Competitive adsorption of heavy metals in aqueous solution onto biochar derived from anaerobically digested sludge. *Chemosphere* **2019**, *219*, 351–357. [CrossRef] [PubMed]
48. Yu, T.; Lv, L.; Wang, H.; Tan, X. Enhanced photocatalytic treatment of Cr(VI) and phenol by monoclinic $BiVO_4$ with {010}-orientation growth. *Mater. Res. Bull.* **2018**, *107*, 248–254. [CrossRef]
49. Du, X.-D.; Deng, J.; Yi, X.-H.; Wang, C.-C.; Zheng, W.; Deng, J.; Wanga, C.C. Robust photocatalytic reduction of Cr(VI) on UiO-66-NH_2(Zr/Hf) metal-organic framework membrane under sunlight irradiation. *Chem. Eng. J.* **2018**, *356*, 393–399. [CrossRef]
50. Abdi, G.; Alizadeh, A.; Zinadini, S.; Moradi, G. Removal of dye and heavy metal ion using a novel synthetic polyethersulfone nanofiltration membrane modified by magnetic graphene oxide/metformin hybrid. *J. Memb. Sci.* **2018**, *552*, 326–335. [CrossRef]
51. Zhang, X.; Fang, X.; Li, J.; Pan, S.; Sun, X.; Shen, J.; Han, W.; Wang, L.; Zhao, S. Developing new adsorptive membrane by modification of support layer with iron oxide microspheres for arsenic removal. *J. Colloid Interface Sci.* **2018**, *514*, 760–768. [CrossRef]

52. Mahdavi, H.R.; Arzani, M.; Isanejad, M.; Mohammadi, T. Effect of hydrophobic and hydrophilic nanoparticles loaded in D2EHPA/M2EHPA-PTFE supported liquid membrane for simultaneous cationic dyes pertraction. *J. Environ. Manag.* **2018**, *213*, 288–296. [CrossRef] [PubMed]
53. Yi, H.; Huang, D.; Qin, L.; Zeng, G.; Lai, C.; Cheng, M.; Ye, S.; Song, B.; Ren, X.; Guo, X. Selective prepared carbon nanomaterials for advanced photocatalytic application in environmental pollutant treatment and hydrogen production. *Appl. Catal. B Environ.* **2018**, *239*, 408–424. [CrossRef]
54. Subramaniam, M.N.; Goh, P.S.; Lau, W.J.; Ng, B.C.; Ismail, A.F. AT-POME colour removal through photocatalytic submerged filtration using antifouling PVDF-TNT nanocomposite membrane. *Sep. Purif. Technol.* **2018**, *191*, 266–275. [CrossRef]
55. Zhang, Q.; Quan, X.; Wang, H.; Chen, S.; Su, Y.; Li, Z. Constructing a visible-light-driven photocatalytic membrane by g-C_3N_4 quantum dots and TiO_2 nanotube array for enhanced water treatment. *Sci. Rep.* **2017**, *7*, 1–7. [CrossRef] [PubMed]
56. Fang, X.; Li, J.; Li, X.; Pan, S.; Zhang, X.; Sun, X.; Shen, J.; Han, W.; Wang, L. Internal pore decoration with polydopamine nanoparticle on polymeric ultrafiltration membrane for enhanced heavy metal removal. *Chem. Eng. J.* **2017**, *314*, 38–49. [CrossRef]
57. Ye, C.; An, Q.; Wu, J.; Zhao, F.; Zheng, P.; Wang, N. Nanofiltration membranes consisting of quaternized polyelectrolyte complex nanoparticles for heavy metal removal. *Chem. Eng. J.* **2019**, *359*, 994–1005. [CrossRef]
58. Yuan, J.; Hung, W.; Zhu, H.; Guan, K.; Ji, Y.; Mao, Y.; Liu, G.; Lee, K.; Jin, W. Fabrication of ZIF-300 membrane and its application for efficient removal of heavy metal ions from wastewater. *J. Memb. Sci.* **2019**, *572*, 20–27. [CrossRef]
59. Attia, H.; Johnson, D.J.; Wright, C.J.; Hilal, N. Comparison between dual-layer (superhydrophobic–hydrophobic) and single superhydrophobic layer electrospun membranes for heavy metal recovery by air-gap membrane distillation. *Desalination* **2018**, *439*, 31–45. [CrossRef]
60. Mahmud, H.N.M.E.; Huq, A.K.O.; Yahya, R. Polymer-based adsorbent for heavy metals removal from aqueous solution. *IOP Conf. Ser. Mater. Sci. Eng.* **2017**, *206*, 012100. [CrossRef]
61. Khulbe, K.C.; Matsuura, T. Removal of heavy metals and pollutants by membrane adsorption techniques. *Appl. Water Sci.* **2018**, *8*, 1–30. [CrossRef]
62. Al-Rashdi, B.A.M.; Johnson, D.J.; Hilal, N. Removal of heavy metal ions by nanofiltration. *Desalination* **2013**, *315*, 2–17. [CrossRef]
63. Hebbar, R.S.; Isloor, A.M.; Ananda, K.; Ismail, A.F. Fabrication of polydopamine functionalized halloysite nanotube/polyetherimide membranes for heavy metal removal. *J. Mater. Chem. A* **2016**, *4*, 764–774. [CrossRef]
64. Ali, S.; Rehman, S.A.U.; Luan, H.Y.; Farid, M.U.; Huang, H. Challenges and opportunities in functional carbon nanotubes for membrane-based water treatment and desalination. *Sci. Total Environ.* **2019**, *646*, 1126–1139. [CrossRef] [PubMed]
65. Mondal, M.; Dutta, M.; De, S. A novel ultrafiltration grade nickel iron oxide doped hollow fiber mixed matrix membrane: Spinning, characterization and application in heavy metal removal. *Sep. Purif. Technol.* **2017**, *188*, 155–166. [CrossRef]
66. Van der Bruggen, B. The Separation power of nanotubes in membranes: A review. *ISRN Nanotechnol.* **2012**, *2012*, 1–17. [CrossRef]
67. Zhang, S.; Liang, S.; Huang, X.; Xiao, K.; Ma, Z.; Wang, H.; Lu, P. A facile approach to fabrication of superhydrophilic ultrafiltration membranes with surface-tailored nanoparticles. *Sep. Purif. Technol.* **2018**, *203*, 251–259.
68. Ai, L.; Zhang, C.; Liao, F.; Wang, Y.; Li, M.; Meng, L.; Jiang, J. Removal of methylene blue from aqueous solution with magnetite loaded multi-wall carbon nanotube: Kinetic, isotherm and mechanism analysis. *J. Hazard. Mater.* **2011**, *198*, 282–290. [CrossRef]
69. Zheng, Y.M.; Zou, S.W.; Nanayakkara, K.G.N.; Matsuura, T.; Chen, J.P. Adsorptive removal of arsenic from aqueous solution by a PVDF/zirconia blend flat sheet membrane. *J. Memb. Sci.* **2011**, *374*, 1–11. [CrossRef]
70. Zahrim, A.Y.; Hilal, N. Treatment of highly concentrated dye solution by coagulation/flocculation-sand filtration and nanofiltration. *Water Resour. Ind.* **2013**, *3*, 23–34. [CrossRef]
71. Pelaez, M.; Nolan, N.T.; Pillai, S.C.; Seery, M.K.; Falaras, P.; Kontos, A.G.; Dunlop, P.S.M.; Hamilton, J.W.J.; Byrne, J.A.; O'Shea, K.; et al. A review on the visible light active titanium dioxide photocatalysts for environmental applications. *Appl. Catal. B Environ.* **2012**, *125*, 331–349. [CrossRef]

72. Nguyen, N.H.; Bai, H. Photocatalytic removal of NO and NO_2 using titania nanotubes synthesized by hydrothermal method. *J. Environ. Sci. (China)* **2014**, *26*, 1180–1187. [CrossRef]
73. Zhang, Y.; Jiang, Z.; Huang, J.; Lim, Y.W.L.; Li, W.; Deng, J.; Gong, D.; Tang, Y.; Lai, Y.; Chen, Z. Titanate and titania nanostructured materials for environmental and energy applications: A review. *RSC Adv.* **2015**, *5*, 79479–79510. [CrossRef]
74. Kijima, T. Inorganic and metallic nanotubular materials: Recent technologies and applications. *Top. Appl. Phys.* **2010**, *117*, 17–32.
75. Gupta, S.M.; Tripathi, M. A review of TiO_2 nanoparticles. *Chin. Sci. Bull.* **2011**, *56*, 1639–1657. [CrossRef]
76. Suarez-Martinez, I.; Grobert, N.; Ewels, C.P. Nomenclature of sp2 carbon nanoforms. *Carbon* **2012**, *50*, 741–747. [CrossRef]
77. Huang, J.; Cao, Y.; Liu, Z.; Deng, Z.; Wang, W. Application of titanate nanoflowers for dye removal: A comparative study with titanate nanotubes and nanowires. *Chem. Eng. J.* **2012**, *191*, 38–44. [CrossRef]
78. Lee, K.; Mazare, A.; Schmuki, P. One-dimensional titanium dioxide nanomaterials: Nanotubes. *Chem. Rev.* **2014**, *114*, 9385–9454. [CrossRef]
79. Yu, J.; Xiang, Q.; Zhou, M. Preparation, characterization and visible-light-driven photocatalytic activity of Fe-doped titania nanorods and first-principles study for electronic structures. *Appl. Catal. B Environ.* **2009**, *90*, 595–602. [CrossRef]
80. Megiel, E.; Ali, G.A.M.; Makhlouf, A.S.H.; Shahryari-ghoshekandi, R.; Nadagouda, M.N.; Gupta, V.K.; Sillanpää, M.; Sadegh, H. The role of nanomaterials as effective adsorbents and their applications in wastewater treatment. *J. Nanostructure Chem.* **2017**, *7*, 1–14.
81. Sadegh, H.; Shahryari-Ghoshekandi, R.; Agarwal, S.; Tyagi, I.; Asif, M.; Gupta, V.K. Microwave-assisted removal of malachite green by carboxylate functionalized multi walled carbon nanotubes: Kinetics and equilibrium study. *J. Mol. Liq.* **2015**, *206*, 151–158. [CrossRef]
82. Aprianti, T.; Miskah, S.; Selpiana; Komala, R.; Hatina, S. Heavy metal ions adsorption from pulp and paper industry wastewater using zeolite/activated carbon-ceramic composite adsorbent. *AIP Conf. Proc.* **2018**, *2014*, 020127.
83. Sharma, M.; Singh, J.; Hazra, S.; Basu, S. Adsorption of heavy metal ions by mesoporous ZnO and TiO_2@ZnO monoliths: Adsorption and kinetic studies. *Microchem. J.* **2019**, *145*, 105–112. [CrossRef]
84. González, F.; Castro, L.; Blázquez, M.L.; Ballester, A.; Muñoz, J.A. Heavy metal adsorption using biogenic iron compounds. *Hydrometallurgy* **2018**, *179*, 44–51.
85. Zhao, W.; Wei, Z.; Ma, L.; Liang, J.; Zhang, X. Ag_2S quantum dots based on flower-like SnS_2 as matrix and enhanced photocatalytic degradation. *Materials* **2019**, *12*, 582. [CrossRef]
86. Li, T.; Wang, Z.; Liu, C.; Tang, C.; Wang, X.; Ding, G.; Ding, Y.; Yang, L. TiO_2 nanotubes/Ag/MoS_2 meshy photoelectrode with excellent photoelectrocatalytic degradation activity for tetracycline hydrochloride. *Nanomaterials* **2018**, *8*, 666. [CrossRef]
87. Wang, Y.; Zhao, W.; Zheng, W.; Chen, S.; Zhao, J. Preparation of N-doped carbon nanosheets from sewage sludge for adsorption studies of Cr(VI) from aqueous solution. *Nanomaterials* **2019**, *9*, 265. [CrossRef]
88. Liang, Y.-C.; Liu, Y.-C. Microstructures and photodegradation performance toward methylene orange of sputtering-assisted decoration of $ZnFe_2O_4$ crystallites onto TiO_2 nanorods. *Nanomaterials* **2019**, *9*, 205. [CrossRef]
89. Jiang, X.F.; Weng, Q.; Wang, X.B.; Li, X.; Zhang, J.; Golberg, D.; Bando, Y. Recent progress on fabrications and applications of boron nitride nanomaterials: A review. *J. Mater. Sci. Technol.* **2015**, *31*, 589–598. [CrossRef]
90. Bernholc, J.; Brenner, D.; Buongiorno Nardelli, M.; Meunier, V.; Roland, C. Mechanical and electrical properties of nanotubes. *Annu. Rev. Mater. Res.* **2002**, *32*, 347–375. [CrossRef]
91. Galstyan, V.; Comini, E.; Faglia, G.; Sberveglieri, G. TiO_2 nanotubes: Recent advances in synthesis and gas sensing properties. *Sensors* **2013**, *13*, 14813–14838. [CrossRef]
92. Bäumer, M.; Weissmüller, J.; Biener, J.; Wittstock, A.; Hamza, A.; Baumann, T. Surface chemistry in nanoscale materials. *Materials* **2009**, *2*, 2404–2428.
93. Shende, P.; Kasture, P.; Gaud, R.S. Nanoflowers: The future trend of nanotechnology for multi-applications. *Artif. Cells Nanomed. Biotechnol.* **2018**, *46*, 413–422. [CrossRef] [PubMed]
94. Bhanjana, G.; Dilbaghi, N.; Kim, K.H.; Kumar, S. Low temperature synthesis of copper oxide nanoflowers for lead removal using sonochemical route. *J. Mol. Liq.* **2017**, *244*, 506–511. [CrossRef]

95. Liu, Y.; Chen, J.; Du, M.; Wang, X.; Ji, X.; He, Z. The preparation of dual-functional hybrid nanoflower and its application in the ultrasensitive detection of disease-related biomarker. *Biosens. Bioelectron.* **2017**, *92*, 68–73. [CrossRef] [PubMed]
96. Zuo, R.; Du, G.; Zhang, W.; Liu, L.; Liu, Y.; Mei, L.; Li, Z. Photocatalytic degradation of methylene blue using TiO_2 Impregnated diatomite. *Adv. Mater. Sci. Eng.* **2014**, *2014*, 170148. [CrossRef]
97. Gangwar, A.; Varghese, S.S.; Meena, S.S.; Prajapat, C.L.; Gupta, N.; Prasad, N.K. Fe_3C nanoparticles for magnetic hyperthermia application. *J. Magn. Magn. Mater.* **2019**, *481*, 251–256. [CrossRef]
98. Huang, Q.; Liu, Y.; Cai, T.; Xia, X. Simultaneous removal of heavy metal ions and organic pollutant by $BiOBr/Ti_3C_2$ nanocomposite. *J. Photochem. Photobiol. A Chem.* **2019**, *375*, 201–208. [CrossRef]
99. Wu, S.; Xiao, L.; Xuan, Y.; Zhao, Z.; Kong, D.; Du, L.; Fang, Z. Research about top electrode improvement of ZnO nanowires array nanogenerator. In Proceedings of the 8th Annual IEEE International Conference on Nano/Micro Engineered and Molecular Systems, Suzhou, China, 7–10 April 2013; pp. 1084–1087.
100. Chen, H.; Shao, L.; Li, Q.; Wang, J. Gold nanorods and their plasmonic properties. *Chem. Soc. Rev.* **2013**, *42*, 2679–2724. [CrossRef] [PubMed]
101. Bobinger, M.; Dergianlis, V.; Becherer, M.; Lugli, P. Comprehensive synthesis study of well-dispersed and solution-processed metal nanowires for transparent heaters. *J. Nanomater.* **2018**, *2018*, 7304807. [CrossRef]
102. Shaban, M.; AbdAllah, H.; Said, L.; Hamdy, H.S.; Abdel Khalek, A. Titanium dioxide nanotubes embedded mixed matrix PES membranes characterization and membrane performance. *Chem. Eng. Res. Des.* **2015**, *95*, 307–316. [CrossRef]
103. Malhotra, A.; Maldovan, M. Thermal transport in semiconductor nanotubes. *Int. J. Heat Mass Transf.* **2019**, *130*, 368–374. [CrossRef]
104. Dehghani, M.H.; Kamalian, S.; Shayeghi, M.; Yousefi, M.; Heidarinejad, Z.; Agarwal, S.; Gupta, V.K. High-performance removal of diazinon pesticide from water using multi-walled carbon nanotubes. *Microchem. J.* **2019**, *145*, 486–491. [CrossRef]
105. Fu, M.; Xing, J.; Ge, Z. Preparation of laccase-loaded magnetic nanoflowers and their recycling for efficient degradation of bisphenol A. *Sci. Total Environ.* **2019**, *651*, 2857–2865. [CrossRef] [PubMed]
106. Jang, H.W.; Nguyen, T.P.; Kim, S.Y.; Jung Park, T.; Hasani, A.; Choi, K.S.; Van Le, Q.; Lee, T.H.; Tekalgne, M. The role of metal dopants in WS_2 nanoflowers in enhancing the hydrogen evolution reaction. *Appl. Catal. A Gen.* **2018**, *567*, 73–79.
107. Runowski, M. Nanotechnology–nanomaterials, nanoparticles and multifunctional core/shell type nanostructures. *Chemik* **2014**, *68*, 766–775.
108. Ealias, A.M.; Saravanakumar, M.P. A review on the classification, characterisation, synthesis of nanoparticles and their application. *IOP Conf. Ser. Mater. Sci. Eng.* **2017**, *263*, 032019.
109. Gusatti, M.; Souza, D.A.R.; Durazzo, M.; Riella, H.G. Chemical processes for the synthesis of nanostructured materials. In *Manufacturing Nanostructures*; One Central Press: Chesire, UK, 2014; pp. 50–78. ISBN 9781910086070.
110. Pareek, V.; Jain, N.; Panwar, J.; Bhargava, A.; Gupta, R. Synthesis and applications of noble metal nanoparticles: A review. *Adv. Sci. Eng. Med.* **2017**, *9*, 527–544. [CrossRef]
111. Wu, W.; Xiao, X.; Zhang, S.; Ren, F.; Jiang, C. Facile method to synthesize magnetic iron oxides/TiO_2 hybrid nanoparticles and their photodegradation application of methylene blue. *Nanoscale Res. Lett.* **2011**, *6*, 533. [CrossRef]
112. Tan, S.; Zhang, X.; Liu, Z.; Gao, J.; Sun, H.; Fu, Y. Hydrothermal synthesis of Ag nanoparticles on the nanocellulose and their antibacterial study. *Inorg. Chem. Commun.* **2018**, *100*, 44–50.
113. Zhang, N.; Chen, D.; Niu, F.; Wang, S.; Qin, L.; Huang, Y. Enhanced visible light photocatalytic activity of Gd-doped $BiFeO_3$ nanoparticles and mechanism insight. *Sci. Rep.* **2016**, *6*, 26467. [CrossRef]
114. Wu, L.; Yang, X.; Li, J.; Huang, Y.; Li, X. Fabrication of titanium dioxide nanotubes with good morphology at high calcination temperature and their photocatalytic activity. *Mater. Chem. Phys.* **2017**, *202*, 136–142. [CrossRef]
115. Alex, S.A.; Chandrasekaran, N.; Mukherjee, A. State-of-the-art strategies for the colorimetric detection of heavy metals using gold nanorods based on aspect ratio reduction. *Anal. Methods* **2016**, *8*, 2131–2137. [CrossRef]
116. Wold, A. Photocatalytic properties of titanium dioxide (TiO_2). *Chem. Mater.* **1993**, *5*, 280–283. [CrossRef]

117. Hashimoto, K.; Irie, H.; Fujishima, A. TiO_2 photocatalysis: A historical overview and future prospects. *Jpn. J. Appl. Phys.* **2005**, *44*, 8269–8285. [CrossRef]
118. Abdelsalam, H.; Teleb, N.H.; Yahia, I.S.; Zahran, H.Y.; Elhaes, H.; Ibrahim, M.A. First principles study of the adsorption of hydrated heavy metals on graphene quantum dots. *J. Phys. Chem. Solids* **2019**, *130*, 32–40. [CrossRef]
119. Wei, S.; Guo, Z.; Bafana, A.; Patel, M.; Guo, J.; Qiu, B.; Lu, Y.; Wujcik, E.K.; Jeffryes, C.; Wang, X.; et al. Carbon nanotubes, graphene, and their derivatives for heavy metal removal. *Adv. Compos. Hybrid Mater.* **2017**, *1*, 56–78.
120. Fu, Y.; Liu, X.; Chen, G. Adsorption of heavy metal sewage on nano-materials such as titanate/TiO_2 added lignin. *Results Phys.* **2019**, *12*, 405–411. [CrossRef]
121. Yarandpour, M.R.; Rashidi, A.; khajavi, R.; Eslahi, N.; Yazdanshenas, M.E. Mesoporous PAA/dextran-polyaniline core-shell nanofibers: Optimization of producing conditions, characterization and heavy metal adsorptions. *J. Taiwan Inst. Chem. Eng.* **2018**, *93*, 566–581. [CrossRef]
122. Wang, Z.; Tan, K.; Cai, J.; Hou, S.; Wang, Y.; Jiang, P.; Liang, M. Silica oxide encapsulated natural zeolite for high efficiency removal of low concentration heavy metals in water. *Colloids Surf. A Physicochem. Eng. Asp.* **2019**, *561*, 388–394. [CrossRef]
123. Liu, J.; Chen, Y.; Han, T.; Cheng, M.; Zhang, W.; Long, J.; Fu, X. A biomimetic SiO_2@chitosan composite as highly-efficient adsorbent for removing heavy metal ions in drinking water. *Chemosphere* **2019**, *214*, 738–742. [CrossRef]
124. Luu, C.L.; Nguyen, Q.T.; Ho, S.T. Synthesis and characterization of Fe-doped TiO_2 photocatalyst by the sol–gel method. *Adv. Nat. Sci. Nanosci. Nanotechnol.* **2010**, *1*, 015008. [CrossRef]
125. Ahmed, M.A.; El-Katori, E.E.; Gharni, Z.H. Photocatalytic degradation of methylene blue dye using Fe_2O_3/TiO_2 nanoparticles prepared by sol–gel method. *J. Alloys Compd.* **2013**, *553*, 19–29. [CrossRef]
126. Aware, D.V.; Jadhav, S.S. Synthesis, characterization and photocatalytic applications of Zn-doped TiO_2 nanoparticles by sol–gel method. *Appl. Nanosci.* **2016**, *6*, 965–972. [CrossRef]
127. Liu, N.; Chen, X.; Zhang, J.; Schwank, J.W. A review on TiO_2-based nanotubes synthesized via hydrothermal method: Formation mechanism, structure modification, and photocatalytic applications. *Catal. Today* **2014**, *225*, 34–51. [CrossRef]
128. Subramaniam, M.N.; Goh, P.S.; Abdullah, N.; Lau, W.J.; Ng, B.C.; Ismail, A.F. Adsorption and photocatalytic degradation of methylene blue using high surface area titanate nanotubes (TNT) synthesized via hydrothermal method. *J. Nanoparticle Res.* **2017**, *19*, 220. [CrossRef]
129. Li, H.B.; Zhang, J.; Huang, G.Y.; Fu, S.H.; Ma, C.; Wang, B.Y.; Huang, Q.R.; Liao, H.W. Hydrothermal synthesis and enhanced photocatalytic activity of hierarchical flower-like Fe-doped $BiVO_4$. *Trans. Nonferrous Met. Soc. China* **2017**, *27*, 868–875. [CrossRef]
130. Huang, M.; Tang, Z.; Yang, J. A new insight for the self-assembly of graphene oxide by hydrothermal method. *Diam. Relat. Mater.* **2019**, *94*, 73–80. [CrossRef]
131. Nikam, A.V.; Prasad, B.L.V.; Kulkarni, A.A. Wet chemical synthesis of metal oxide nanoparticles: A review. *CrystEngComm* **2018**, *20*, 5091–5107. [CrossRef]
132. Zhang, W.; Zhang, Z.; Kwon, S.; Zhang, F.; Stephen, B.; Kim, K.K.; Jung, R.; Kwon, S.; Chung, K.B.; Yang, W. Photocatalytic improvement of Mn-adsorbed g-C_3N_4. *Appl. Catal. B Environ.* **2017**, *206*, 271–281. [CrossRef]
133. Miller, C.J.; Yu, H.; Waite, T.D. Degradation of rhodamine B during visible light photocatalysis employing Ag@AgCl embedded on reduced graphene oxide. *Colloids Surf. A Physicochem. Eng. Asp.* **2013**, *435*, 147–153. [CrossRef]
134. Zhong, Y.; Li, H.; Xu, Q.; He, J.; Lu, J.; Chen, D.; Dong, R.; Li, N. Morphology-controlled fabrication of CNT@MoS_2/SnS_2 nanotubes for promoting photocatalytic reduction of aqueous Cr(VI) under visible light. *J. Alloys Compd.* **2019**, *784*, 282–292.
135. Chandra, R. Synthesis of mixed metal oxide nanoparticles of SnO_2 with SrO via sol–gel technology: Their structural, optical, and electrical properties. *J. Sol-Gel Sci. Technol.* **2018**, *87*, 41–49.
136. Lu, D.; Kumar Kondamareddy, K.; Fan, H.; Gao, B.; Wang, J.; Wang, J.; Hao, H. Highly improved visible-light-driven photocatalytic removal of Cr(VI) over yttrium doped H-Titanate nanosheets and its synergy with organic pollutant oxidation. *Sep. Purif. Technol.* **2019**, *210*, 775–785. [CrossRef]
137. Bayal, N.; Jeevanandam, P. Synthesis of TiO_2-MgO mixed metal oxide nanoparticles via a sol-gel method and studies on their optical properties. *Ceram. Int.* **2014**, *40*, 15463–15477. [CrossRef]

138. Mousavi, S.J.; Parvini, M.; Ghorbani, M. Adsorption of heavy metals (Cu^{2+} and Zn^{2+}) on novel bifunctional ordered mesoporous silica: Optimization by response surface methodology. *J. Taiwan Inst. Chem. Eng.* **2018**, *84*, 123–141. [CrossRef]
139. Deng, J.; You, Y.; Sahajwalla, V.; Joshi, R.K. Transforming waste into carbon-based nanomaterials. *Carbon* **2016**, *96*, 105–115. [CrossRef]
140. Kumar, P.S.; Venkatesh, K.; Gui, E.L.; Jayaraman, S.; Singh, G.; Arthanareeswaran, G. Electrospun carbon nanofibers/TiO_2-PAN hybrid membranes for effective removal of metal ions and cationic dye. *Environ. Nanotechnol. Monit. Manag.* **2018**, *10*, 366–376. [CrossRef]
141. Overah, L.C.; Iwegbue, C.M.; Babalola, J.O.; Martincigh, B.S. Fabrication and characterisation of a Fe_3O_4/Raphia farinifera nanocomposite for application in heavy metal adsorption. *Environ. Technol. Innov.* **2019**, *13*, 11–29. [CrossRef]
142. Chen, J.; Zhu, J.; Wang, N.; Feng, J.; Yan, W. Hydrophilic polythiophene/SiO_2 composite for adsorption engineering: Green synthesis in aqueous medium and its synergistic and specific adsorption for heavy metals from wastewater. *Chem. Eng. J.* **2018**, *360*, 1486–1497. [CrossRef]
143. Sherlala, A.I.A.; Raman, A.A.A.; Bello, M.M.; Asghar, A. A review of the applications of organo-functionalized magnetic graphene oxide nanocomposites for heavy metal adsorption. *Chemosphere* **2018**, *193*, 1004–1017. [CrossRef]
144. Marciniak, M.; Goscianska, J.; Frankowski, M.; Pietrzak, R. Optimal synthesis of oxidized mesoporous carbons for the adsorption of heavy metal ions. *J. Mol. Liq.* **2019**, *276*, 630–637. [CrossRef]
145. Li, L.Y.; Gong, X.D.; Abida, O. Waste-to-resources: Exploratory surface modification of sludge-based activated carbon by nitric acid for heavy metal adsorption. *Waste Manag.* **2019**, *87*, 375–386. [CrossRef]
146. Luo, S.; Cai, T.; Liu, C.; Zhang, Y.; Liu, Y.; Ma, J.; Wei, Y.; Ali, O.; Zhang, S. Fast adsorption of heavy metal ions by waste cotton fabrics based double network hydrogel and influencing factors insight. *J. Hazard. Mater.* **2017**, *344*, 1034–1042.
147. Wang, S.; Ning, H.; Hu, N.; Huang, K.; Weng, S.; Wu, X.; Wu, L.; Liu, J. Alamusi preparation and characterization of graphene oxide/silk fibroin hybrid aerogel for dye and heavy metal adsorption. *Compos. Part B Eng.* **2019**, *163*, 716–722. [CrossRef]
148. Siyal, A.A.; Shamsuddin, M.R.; Khan, M.I.; Rabat, N.E.; Zulfiqar, M.; Man, Z.; Siame, J.; Azizli, K.A. A review on geopolymers as emerging materials for the adsorption of heavy metals and dyes. *J. Environ. Manage.* **2018**, *224*, 327–339. [CrossRef]
149. Ghiloufi, I.; El Ghoul, J.; Modwi, A.; Mir, L. El Ga-doped ZnO for adsorption of heavy metals from aqueous solution. *Mater. Sci. Semicond. Process.* **2016**, *42*, 102–106. [CrossRef]
150. Gong, J.; Wang, X.; Shao, X.; Yuan, S.; Yang, C.; Hu, X. Adsorption of heavy metal ions by hierarchically structured magnetite-carbonaceous spheres. *Talanta* **2012**, *101*, 45–52. [CrossRef] [PubMed]
151. Kar, P.; Jain, P.; Kumar, V.; Gupta, R.K. Interfacial engineering of Fe_2O_3@BOC heterojunction for efficient detoxification of toxic metal and dye under visible light illumination. *J. Environ. Chem. Eng.* **2019**, *7*, 102843. [CrossRef]
152. Wang, P.; Yi, X.; Du, X.; Wang, C.; Deng, J. Enhanced photocatalytic Cr(VI) reduction and diclofenac sodium degradation under simulated sunlight irradiation over MIL-100(Fe)/g-C_3N_4 heterojunctions. *Chinese J. Catal.* **2018**, *40*, 70–79. [CrossRef]
153. Kumar, K.V.A.; Chandana, L.; Ghosal, P.; Subrahmanyam, C. Simultaneous photocatalytic degradation of p-cresol and Cr(VI) by metal oxides supported reduced graphene oxide. *Mol. Catal.* **2018**, *451*, 87–95. [CrossRef]
154. Koh, P.W.; Hatta, M.H.M.; Ong, S.T.; Yuliati, L.; Lee, S.L. Photocatalytic degradation of photosensitizing and non-photosensitizing dyes over chromium doped titania photocatalysts under visible light. *J. Photochem. Photobiol. A Chem.* **2017**, *332*, 215–223. [CrossRef]
155. Liu, E.; Du, Y.; Bai, X.; Fan, J.; Hu, X. Synergistic improvement of Cr(VI) reduction and RhB degradation using RP/g-C_3N_4 photocatalyst under visible light irradiation. *Arab. J. Chem.* **2019**. [CrossRef]
156. Ruan, X.; Hu, H.; Che, H.; Jiang, E.; Zhang, X.; Liu, C.; Che, G. A visible-light-driven Z-scheme CdS/$Bi_{12}GeO_{20}$ heterostructure with enhanced photocatalytic degradation of various organics and the reduction of aqueous Cr(VI). *J. Colloid Interface Sci.* **2019**, *543*, 317–327. [CrossRef] [PubMed]

157. Ye, J.; Cui, L.; Huang, Z.; Huang, Z.; Wu, P.; Dai, X.; Wang, T. Simultaneous and efficient photocatalytic reduction of Cr(VI) and oxidation of trace sulfamethoxazole under LED light by rGO@Cu_2O/$BiVO_4$ p-n heterojunction composite. *Chemosphere* **2019**, *221*, 824–833.
158. Xu, S.; Dai, J.; Pang, X.; Yang, J.; Wang, Y.; Hao, J. Visible-light-driven photocatalytic degradation of 4-CP and the synergistic reduction of Cr(VI) on one-pot synthesized amorphous Nb_2O_5 nanorods/graphene heterostructured composites. *Chem. Eng. J.* **2018**, *353*, 100–114.
159. Kaur, K.; Jindal, R. Synergistic effect of organic-inorganic hybrid nanocomposite ion exchanger on photocatalytic degradation of Rhodamine-B dye and heavy metal ion removal from industrial effluents. *J. Environ. Chem. Eng.* **2018**, *6*, 7091–7101. [CrossRef]
160. Ghafoor, S.; Hussain, S.Z.; Waseem, S.; Arshad, S.N. Photo-reduction of heavy metal ions and photo-disinfection of pathogenic bacteria under simulated solar light using photosensitized TiO_2 nanofibers. *RSC Adv.* **2018**, *8*, 20354–20362. [CrossRef]
161. Bai, X.; Du, Y.; Hu, X.; He, Y.; He, C.; Liu, E.; Fan, J. Synergy removal of Cr(VI) and organic pollutants over RP-MoS_2/rGO photocatalyst. *Appl. Catal. B Environ.* **2018**, *239*, 204–213. [CrossRef]
162. Ye, M.; Wei, W.; Zheng, L.; Liu, Y.; Wu, D.; Gu, X.; Wei, A. Enhanced visible light photoreduction of aqueous Cr(VI) by Ag/Bi_4O_7/g-C_3N_4 nanosheets ternary metal/non-metal Z-scheme heterojunction. *J. Hazard. Mater.* **2019**, *365*, 674–683. [CrossRef]
163. Zhou, M.; Yang, P.; Wang, S.; Luo, Z.; Huang, C.; Wang, X. Structure-mediated charge separation in boron carbon nitride for enhanced photocatalytic oxidation of alcohol. *ChemSusChem* **2018**, *11*, 3949–3955. [CrossRef]
164. Shinde, S.S.; Kher, J.A. A review on polyaniline and its noble metal composites. *Int. J. Innov. Res. Sci. Eng. Technol.* **2014**, *3*, 16570–16576. [CrossRef]
165. Xie, M.; Huang, L.; Liu, J.; Xu, Y.; Deng, J.; Li, H.; Xu, H.; Jing, L. Three dimensional polyaniline/$MgIn_2S_4$ nanoflower photocatalysts accelerated interfacial charge transfer for the photoreduction of Cr(VI), photodegradation of organic pollution and photocatalytic H2 production. *Chem. Eng. J.* **2018**, *360*, 1601–1612.
166. Wang, G.; Fan, W.; Li, Q.; Deng, N. Enhanced photocatalytic new coccine degradation and Pb(II) reduction over graphene oxide-TiO_2 composite in the presence of aspartic acid-B-cyclodextrin. *Chemosphere* **2019**, *216*, 707–714. [CrossRef] [PubMed]
167. Dadigala, R.; Bandi, R.; Gangapuram, B.R.; Dasari, A.; Belay, H.H.; Guttena, V. Fabrication of novel 1D/2D V_2O_5/g-C_3N_4 composites as Z-scheme photocatalysts for CR degradation and Cr(VI) reduction under sunlight irradiation. *J. Environ. Chem. Eng.* **2019**, *7*, 102822. [CrossRef]
168. Barati, R.; Gilan, N.; Yousefi, N.; Ghasemi, S.; Ahmadian, M.; Moussavi, S.; Rahimi, S.; Fatehizadeh, A.; Rahimi, K.; Reshadat, S. Photocatalytic removal of cadmium (II) and lead (II) from simulated wastewater at continuous and batch system. *Int. J. Environ. Health Eng.* **2014**, *3*, 31.
169. Wang, F.X.; Yi, X.H.; Wang, C.C.; Deng, J.G. Photocatalytic Cr(VI) reduction and organic-pollutant degradation in a sTable 2D coordination polymer. *Cuihua Xuebao/Chinese J. Catal.* **2017**, *38*, 2141–2149. [CrossRef]
170. Deng, F.; Lu, X.; Luo, Y.; Wang, J.; Che, W.; Yang, R.; Luo, X.; Luo, S.; Dionysiou, D.D. Novel visible-light-driven direct Z-scheme CdS/$CuInS_2$ nanoplates for excellent photocatalytic degradation performance and highly-efficient Cr(VI) reduction. *Chem. Eng. J.* **2019**, *361*, 1451–1461. [CrossRef]
171. Kumordzi, G.; Malekshoar, G.; Yanful, E.K.; Ray, A.K. Solar photocatalytic degradation of Zn^{2+} using graphene based TiO_2. *Sep. Purif. Technol.* **2016**, *168*, 294–301. [CrossRef]
172. Priyadharsan, A.; Vasanthakumar, V.; Karthikeyan, S.; Raj, V.; Shanavas, S.; Anbarasan, P.M. Multi-functional properties of ternary CeO_2/SnO_2/rGO nanocomposites: Visible light driven photocatalyst and heavy metal removal. *J. Photochem. Photobiol. A Chem.* **2017**, *346*, 32–45. [CrossRef]
173. Sunil, K.; Karunakaran, G.; Yadav, S.; Padaki, M. Al-Ti_2O_6 a mixed metal oxide based composite membrane: A unique membrane for removal of heavy metals. *Chem. Eng. J.* **2018**, *348*, 678–684. [CrossRef]
174. Ghaemi, N.; Daraei, P. Enhancement in copper ion removal by PPy@Al_2O_3 polymeric nanocomposite membrane. *J. Ind. Eng. Chem.* **2016**, *40*, 26–33. [CrossRef]
175. Ghaemi, N.; Zereshki, S.; Heidari, S. Removal of lead ions from water using PES-based nanocomposite membrane incorporated with polyaniline modified GO nanoparticles: Performance optimization by central composite design. *Process Saf. Environ. Prot.* **2017**, *111*, 475–490. [CrossRef]

176. Anjum, M.; Miandad, R.; Waqas, M.; Gehany, F.; Barakat, M.A. Remediation of wastewater using various nano-materials. *Arab. J. Chem.* **2016**. [CrossRef]
177. He, J.; Matsuura, T.; Chen, J.P. A novel Zr-based nanoparticle-embedded PSF blend hollow fiber membrane for treatment of arsenate contaminated water: Material development, adsorption and filtration studies, and characterization. *J. Memb. Sci.* **2014**, *452*, 433–445. [CrossRef]
178. Zhao, D.; Yu, Y.; Chen, J.P. Treatment of lead contaminated water by a PVDF membrane that is modified by zirconium, phosphate and PVA. *Water Res.* **2016**, *101*, 564–573. [CrossRef] [PubMed]
179. Delavar, M.; Bakeri, G.; Hosseini, M. Fabrication of polycarbonate mixed matrix membranes containing hydrous manganese oxide and alumina nanoparticles for heavy metal decontamination: Characterization and comparative study. *Chem. Eng. Res. Des.* **2017**, *120*, 240–253. [CrossRef]
180. Jamshidi Gohari, R.; Lau, W.J.; Matsuura, T.; Halakoo, E.; Ismail, A.F. Adsorptive removal of Pb(II) from aqueous solution by novel PES/HMO ultrafiltration mixed matrix membrane. *Sep. Purif. Technol.* **2013**, *120*, 59–68. [CrossRef]
181. Li, L.; Wang, F.; Lv, Y.; Liu, J.; Zhang, D.; Shao, Z. Halloysite nanotubes and Fe_3O_4 nanoparticles enhanced adsorption removal of heavy metal using electrospun membranes. *Appl. Clay Sci.* **2018**, *161*, 225–234. [CrossRef]
182. Khadijah, S.; Ha, M.; Othman, D.; Harun, Z.; Ismail, A.F.; Rahman, M.A.; Jaafar, J. A novel green ceramic hollow fiber membrane (CHFM) derived from rice husk ash as combined adsorbent-separator for efficient heavy metals removal. *Ceram. Int.* **2017**, *43*, 4716–4720.
183. Ali, S.; Aziz, S.; Rehman, U.; Ali, I.; Usman, M. Efficient removal of zinc from water and wastewater effluents by hydroxylated and carboxylated carbon nanotube membranes: Behaviors and mechanisms of dynamic filtration. *J. Hazard. Mater.* **2019**, *365*, 64–73. [CrossRef]
184. Attia, H.; Alexander, S.; Wright, C.J.; Hilal, N. Superhydrophobic electrospun membrane for heavy metals removal by air gap membrane distillation (AGMD). *Desalination* **2017**, *420*, 318–329. [CrossRef]
185. Zhang, Y.; Zhang, S.; Gao, J.; Chung, T. Layer-by-layer construction of graphene oxide (GO) framework composite membranes for highly efficient heavy metal removal. *J. Memb. Sci.* **2016**, *515*, 230–237. [CrossRef]
186. Lee, C.; Chiang, C.; Liu, S. Electrospun nanofibrous rhodanine/polymethylmethacrylate membranes for the removal of heavy metal ions. *Sep. Purif. Technol.* **2013**, *118*, 737–743. [CrossRef]
187. Soghra, S.; Zahed, H.; Khanlari, S.; Mohammadi, T. Hydrous metal oxide incorporated polyacrylonitrile-based nanocomposite membranes for Cu(II) ions removal. *Sep. Purif. Technol.* **2019**, *213*, 151–161.
188. Sevinc, P.C.; Dhital, B.; Govind Rao, V.; Wang, Y.; Lu, H.P. Probing electric field effect on covalent interactions at a molecule-semiconductor interface. *J. Am. Chem. Soc.* **2016**, *138*, 1536–1542. [CrossRef]
189. Chavez, S.; Rao, V.G.; Linic, S. Unearthing the factors governing site specific rates of electronic excitations in multicomponent plasmonic systems and catalysts. *Faraday Discuss.* **2019**. [CrossRef]

© 2019 by the authors. Licensee MDPI, Basel, Switzerland. This article is an open access article distributed under the terms and conditions of the Creative Commons Attribution (CC BY) license (http://creativecommons.org/licenses/by/4.0/).

nanomaterials

MDPI

Review

In-Situ Synthesized Si@C Materials for the Lithium Ion Battery: A Mini Review

Wenmao Tu, Ziyu Bai, Zhao Deng, Haining Zhang and Haolin Tang *

State Key Laboratory of Advanced Technology for Materials Synthesis and Processing, Wuhan University of Technology, Wuhan 430070, China; tuwm@whut.edu.cn (W.T.); 18971142560@163.com (Z.B.); dengzhao@whut.edu.cn (Z.D.); Haining.zhang@whut.edu.cn (H.Z.)

* Correspondence: thln@whut.edu.cn; Tel.: +86-27-8788-4448

Received: 27 January 2019; Accepted: 7 March 2019; Published: 14 March 2019

Abstract: As an important component, the anode determines the property and development of lithium ion batteries. The synthetic method and the structure design of the negative electrode materials play decisive roles in improving the property of the thus-assembled batteries. Si@C compound materials have been widely used based on their excellent lithium ion intercalation capacity and cyclic stability, in which the in-situ synthetic method can make full use of the structural advantages of the monomer itself, thus improving the electrochemical performance of the anode material. In this paper, the different preparation technologies and composite structures of Si@C compound materials by in-situ synthesis are introduced. The research progress of Si@C compound materials by in-situ synthesis is reviewed, and the prospect of future development of Si@C compound materials has been tentatively commented.

Keywords: lithium ion battery; Si@C compound material; in-situ synthesis

1. Introduction

The lithium ion battery is one of the ideal green rechargeable energy conversion devices. Its principle is based on the lithium ion intercalation and deintercalation between a carbon negative electrode and a compound positive electrode. Lithium ion batteries have been widely used in portable electronic products because of their advantages, including large energy density, small self-discharge, high output voltage, and good security [1,2]. The commercially available anode materials for lithium ion batteries are mainly graphite and other carbonaceous materials. However, the poor rate performance indicates that it is difficult to meet the demand of downstream products of lithium-ion batteries. In the field of consumer electronics, the energy density of the battery needs to be improved. The Si@C compound material, which is a new kind of high-capacity negative materials, will become the development trend of lithium ion batteries.

The main advantages of silicon-based materials are depicted as follows. (1) The excellent capacity [3]; (2) the relatively stable amorphous microstructure after the first lithium intercalation; (3) difficulty in reunion of materials during the process of lithium ion deintercalation. In addition, it is also not easy to generate the lithium dendrite, because of the higher discharge platform compared to carbon-based materials, and there is no co-embedding of electrolytes. The main disadvantage is the serious volume effect during the process of high lithium ion deintercalation, resulting in the collapse of the electrode materials and the exfoliation of active substances, characterizing a rapid capacity attenuation [4]. Based on the advantages and disadvantages of silicon materials and carbon materials, it is a good choice to synthesize Si@C compound materials, with an optimized design of structure, as the negative material [5]. In such a system, silicon is used as an active substance to provide lithium storage capacity [6]; carbon is used as a dispersed matrix to buffer the volume effect of silicon particles in the process of charging–discharging, in order to maintain the electrical contact within the electrode and the integrity of the electrode structure [7]. The preparation of Si@C anode materials by using the in-situ synthesis method can make full use of the structural advantages of the monomer, and creates a good fusion of the advantages of their respective properties, thus improving the performance of the composites.

Numerous lithium-ion Si@C anode materials have been designed to buffer the volume expansion of silicon and to optimize the lithium intercalation performance of silicon, by controlling synthetic methods and designing different structural models. For example, Si@C composites are prepared by efficient mixing using high-energy ball-milling, or by adding the eff to link the silicon-containing materials and carbon-containing materials, so as to achieve different structural designs for improvement in electrochemical properties. Aiming at in-situ synthesis, this method is developed on the basis of enhancing the mechanical and physical active surfactant properties of inorganic materials. Based on the advantages of in-situ synthesis, it has been extensively applied to organic synthesis and adsorption processes in environmental engineering, but it is seldom used in energy storage. By analyzing the properties of the required composites, suitable in-situ monomers are selected to retain the performance advantages of the monomers in the system, and the structural advantages of the monomers can also be designed. For Si@C anode materials of lithium-ion batteries, high performance anode materials can be prepared by in-situ electrochemical synthesis using alloying products during charging and discharging, and solid-phase in-situ synthesis can also be carried out on the basis of raw material monomers. The in-situ synthesis method based on environmental protection adsorbent materials is applied to prepare the components of lithium storage devices, which is a novel development direction of negative electrode composite for lithium ion batteries.

2. In-Situ Synthesis of Si@C Compound Materials

Generally, the synthetic process which is based on a hybrid precursor comprising both silicon components and organic components, and involves simultaneous formation of active electrode materials and carbonaceous protective coating is referred to an "in-situ synthesis". In-situ synthesis is a recently developed technology for preparing compound materials. The basic principle relies on the occurred chemical reaction between different elements or compounds under a certain condition and the generated secondary phases in the substrate phase to improve the performance of single substrate. In such a compound system, reinforcement spontaneously grows in the substrate phase and the interfacial bonding strength can be improved. For in-situ synthesis, if there are organic compounds in the synthetic monomers, organic–inorganic polymers are often formed. Inorganic molecules can be grafted on the organic monomers to form a homogeneous composite of an encapsulated type, rather than a simple physical compound formed by mechanical mixing. If the monomers used for synthesis are inorganic, the co-embedding of monomers can be realized. The bonding strength is obviously greater than the physical force of mechanical mixing, so that the properties of the composite will fully combine the advantages of monomer properties. Compared with ex-situ synthesis, the structure of the composites synthesized is more uniform, and the design of specific structures can be achieved by the selection of monomers.

2.1. *In-Situ Electrochemical Synthesis*

In the process of alloying and dealloying of lithium ion with Si, silicon-based materials will cause severe structural damage due to its huge volume effect. Thus, the specific capacity is poor owing to the loss of electrical contact between the cracked and isolated Si particles [8,9]. However, based on the formation of a high lithium intercalation intermetallic compound and the appearance of amorphous phase in the material system, the silicon-based materials have high theoretical specific capacity. Therefore, how to reduce the shortcomings of silicon-based anode materials, reasonably, and fully tap their theoretical specific capacity will be the future research focus for silicon-based anode materials. To solve the problem of structural instability of silicon-based anodes, the Si@C composite by in-situ electrochemical synthesis has been developed. The in-situ method can generate the active–inactive product in which the silicon is evenly dispersed in the carbon matrix, to improve the mechanical properties and the electronic conductivity of materials [10–12]. This inactive or relatively active matrix exhibits good mechanical strength to buffer the stress and strain generated by the charge and discharge process of the active silicon phase. In addition, the substrate with electrochemical active phase must have high conductivity and lithium ion mobility to make the material possess a high rate performance.

At the initial stage of the charge and discharge process, the electrochemical cycle is limited in the Li-Si intermetallic compound, rather than in the silicon particles, which would help to alleviate the volume effect during the alloying process. Moni et al. [13] reported the generation of lithialized Si@C compound material by in-situ electrochemical synthesis using high energy mechanical milling (HEMM). The theoretical specific capacity of the Li-Si@C composite based on different Li-Si alloy compositions is shown in Table 1. In this report, multiple contrasting samples of Li-Si@C were prepared by mixing the Si@C composite prepared by HEMM and PAN (polyaniline) with 42 wt.% Gr, 28 wt.% Si, and 30 wt.% PAN under various electrochemical reaction conditions and different components. The materials were obtained at ~0.6 V, ~0.5 V, and ~0.4 V, with the Li-Si alloy dispersed in graphite matrix. Under constantly cycling at stable potential, the composite exhibited better cyclability in comparison to the pure Si@C composite. The energy density was improved because of the higher discharge potential for the full cell. The Li-Si@C compound material with 64 at.%C, 21.6 at.% Li, and 14.4 at.% Si could have a stable cycle in the potential window of 0.02–0.5 V, which is similar to the composite with Li-40 at.% Si dispersed in the carbon matrix, and the composite obtains a high energy density in the full cell configuration with excellent stability (~0.13% loss per cycle and capacity of 700 $mAhg^{-1}$).

2.2. In-Situ Solid-State Synthesis

Yushin et al. reported on the coated nano silicon@graphene composite particle that is prepared by the method of a layered, bottom-up assembly, through continuous silicon and carbon chemical vapor deposition (CVD) processes, started from multilayer graphene gel [14,15]. After accurately creating and adjusting nano-scale pores in self-assembled particles, the three-dimensional porous particle structure, consisting of a curved two-dimensional layer, can be used as a buffer to adapt to the volume expansion of silicon and provide irregular channels for rapid acquisition of lithium ion transport, thus improving the reversible capacity, columbic efficiency, and cyclability. Although the CVD process using high purity propylene and highly flammable silane have a high cost, which may impede its commercial applications, this bottom-up strategy can be used to prepare hierarchical complex substances that have a long cycle life. Therefore, the direct synthesis of graphene-based composites by in-situ growth of graphene is considered as a new important strategy.

Zhang et al. [16] reported the synthesis of silicon@graphitic carbon via a one-step in-situ solid-state reaction. The schematic diagram of the formation of the N-doped porous graphene frame-supported Si@graphite carbon and the cycling performance of SiNPs, Graphene, and N-doped graphene frame-supported Si@graphite carbon (NGSi@G) granules are shown in Figure 1. In this system, the in-situ-formed nitrogen-doped graphene wrapped the SiNPs through thermal decomposition and re-carbonization of FePC, forming a cross-linked graphene skeleton network, and SiNP was encapsulated by a thin graphite carbon layer. The self-assembly process occurred based on the strong interconnections between the graphene, graphite carbon, and SiNPs, resulting in the formation of nanoparticles with tunable nano-porosities and irregular channels. The self-assembled particles with good conductivity can effectively adapt to the volume change of SiNPs during the reduplicate charge–discharge process, thus achieving high capacity, long cycle life, high coulomb efficiency, and excellent rate capacity. The results show that the graphene frame-supported silicon@graphitic carbon granules have high reversible capacities of 1345 and 1065 $mAhg^{-1}$ after 100 and 200 cycles, respectively, corresponding to capacity retentions of 91% and 72%, respectively. A high reversible capacity of 1042 $mAhg^{-1}$ at 28 Ag^{-1} (10C) can also be achieved, revealing a superb rate capability.

Table 1. The theoretical specific capacity with Li-Si–C composite based on different Li-Si alloy compositions.

Alloys	Composition	Phases	Specific Capacity ($mAhg^{-1}$)	Capacity Loss per Cycle (%)	Charge Potential (V)
Si/C	82.6 at.% C, 18.4 at.% Si	Si + C	1000	0.34	1.2
Li-Si/C(0.6 V)	69 at.% C, 15.5 at.% Si, 15.5 at.% Li	Li-Si + C	800	0.21	0.6
Li-Si/C(0.5 V)	64 at.% C, 21.6 at.% Si, 14.4 at.% Li	Li-Si + Li7Si3 + C	700	0.13	0.5
Li-Si/C(0.4 V)	57.1 at.% C, 30 at.% Si, 12.9 at.% Li	Li7Si3 + C	440	3.6	0.4
C	-	C	320	0%	1.2
Si	-	Si	4000	-	1.2

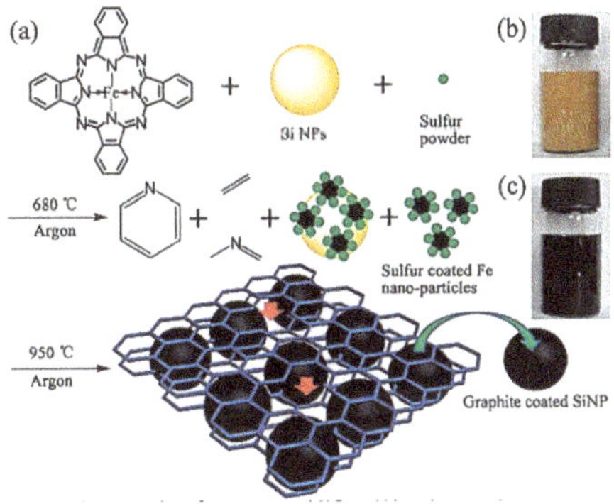

Figure 1. (**a**) Schematic diagram of the formation of the N-doped porous graphene frame-supported Si@graphite carbon granules; (**b**,**c**) images of SiNPs and as-prepared granules dispersed in ethanol solution.

2.3. In-Situ Carbothermal Reduction

Carbon thermal in-situ synthesis has its unique advantages: (1) The decomposition products of high molecular polymers have higher reducibility than solid carbon, which can reduce the synthesis temperature and shorten the reaction time; (2) the derived carbon from polymer are atomically dispersed in the reaction system, leading to the homogeneous coating of synthetic products; (3) in-situ-coated carbon films can reduce the growth rate of $LiFePO_4$. By using in-situ synthesis depending on chemical reaction, nanoscale silicon can be generated and it will be well-dispersed in the buffer matrix to form the composite. The prepared composite can suppress the absolute volume change and the cyclic stability of the material system can be improved accordingly [12,17,18]. Zhang et al. used the carbon thermal in-situ reduction method to prepare nano silicon (less than 50 nm) that was dispersed in a carbon matrix [19]. Under inert atmosphere, the Si@C composite was prepared by milling silicon oxide particles and sucrose, and sintering in a certain temperature. The composite with 20% Si, 80% C still maintained a high reversible capacity after 100 cycles, with an average attenuation rate of only 0.27%. The improvement in the cyclic performance of the Si@C compound material is mainly attributed to the effective buffer of the volume effect and the improved conductivity of graphite materials.

Rice husk is a natural material in which silicon and carbon sources coexist. The silicon dioxide is mostly concentrated on the upper and lower epidermis to form a compact silica shell, and a small amount of silica is distributed in its internal carbon material. The carbon materials, represented by lignin and cellulose, are mainly distributed in the internal vascular bundle [20]. Wang et al. carried out the pyrolysis of rice husk, graphitization at 1000 °C, and removed silicon dioxide to obtain rice hull-based carbon fibers. The material showed good capacity retention and rate capability [21]. Bao et al. obtained porous silicon from rice husk by use of the magnesium heat reduction method; the specific capacity could still reach 1220 $mAhg^{-1}$ after 100 cycles at the current density of 1 Ag^{-1} and it also had a good rate performance [22]. Based on the preparation of rice husk silicon by magnesium heat reduction, Jung et al. prepared the Si@C compound material with a coated layer structure using poly(dopamine) as precursor. This material exhibits good electrochemical performance. The specific capacity of the first 200 cycles is above 1500 $mAhg^{-1}$, and the rate capacity is good [23]. Wang et al. used alkali boiled in-situ carbothermal reduction to synthesize the Si@C compound anode material. In this method, the rice husk is pretreated with sodium hydroxide. After carbonization and magnesium heat reduction, the Si@C compound anode material was obtained. The composite has excellent rate capacity [24]. Si@C compound anode material can be also prepared by in-situ synthesis of different ash template carbons.

3. The Structure and Electrochemical Properties of Si@C Compound Materials

For the coated composite, the surface carbon layer is mainly amorphous carbon [14], and for the embedded composite, the surface carbon layer can be graphite [25,26] or graphene. The three-component composite structure between silicon and carbon is also a hot topic at present. The synthetic method and electrochemical performance comparison of Si@C-based anode materials in the lithium ion battery are shown in Table 2.

3.1. Coated Composite

In the coated silicon–carbon composites, the high silicon content contributes to the increase in the lithium storage capacity, and the carbon layer coated on the surface of silicon can effectively alleviate the volume effect of silicon and enhance the electronic conductivity. For this system, a stable solid electrolyte membrane (SEI) can be produced to stabilize the interface between the electrode material and the electrolyte [27]. In order to solve the possible fracture of the carbon layer caused by intense silicon stress, research works to optimize the microstructure of the carbon layer and to prepare the nanofiber composite have been published [28].

Gao et al. used in-situ polymerization and pyrolysis to prepare a Si@C compound material that had a core-shell structure, and its surface coating layer had a complete microporous structure. The negative electrode still maintained the reversible capacity of 1200 $mAhg^{-1}$ after 40 cycles [29]. Cui et al. used the CVD method to deposit silicon on the surface of carbon fiber to prepare the nanowire Si@C compound material with a core-shell structure. After 30 cycles at the ratio of 0.2 C, the lithium storage capacity up to 2000 $mAhg^{-1}$ was maintained [28]. Wang et al. [30] prepared a nano-silicon@C composite with a core-shell structure using a ball-milling and carbonization approach. In contrast to micro-silicon, the system with nano-silicon had a higher specific capacity. As a coating layer, carbon can make use of the buffering effect to improve the capacity retention. After 100 cycles, the specific capacity can be maintained at 1060 $mAhg^{-1}$. It is well known that the existence of a core-shell structure can effectively solve the volume effect of silicon and improve the electrochemical performance. Zhou et al [31] synthesized Si/graphite@N-doped carbon with a core-shell structure and doping system using a two-step synthesis, including liquid mixing and spray-drying, to prepare silicon–graphite particles and oxidative self-polymerization of dopamine to compare coating layer. The composite exhibited a high specific capacity of 611.3 $mAhg^{-1}$ after 100 cycles. As a development of the solid Si@C core-shell structure, the introduction of additional internal void spaces into the core-shell structure to form a yolk-shell structure can make particles contact better. The voids in this structure can buffer the volume expansion of silicon and allow the core to shrink without pulverization, which is beneficial to form stable SEI and to hold the integrity of the electrode. Yang et al [32] used a template-based and nanocasting method to synthesize a yolk-shell silicon-mesoporous carbon material. The schematic illustration and electrochemical tests are shown in Figure 2. In this structure, single silicon nanoparticles are encapsulated in an open and accessible mesoporous carbon layer rather than a solid carbon layer. In addition to providing enough space for volume expansion, the porosities of carbon shells allow rapid transport of lithium ions between electrolytes and silicon egg yolks. It was presented that the specific capacity was maintained at 1000 $mAhg^{-1}$ after 400 cycles and the superior rate capability of 62.3% capacity retention at a current density of 8.4 Ag^{-1} was realized.

Table 2. Synthesis method and electrochemical performance comparison of Si@C-based anode materials in the lithium ion battery.

Anodes	Specific Capacity ($mAhg^{-1}$)	Cycling Stability ($mAhg^{-1}$)	Rate Capacity ($mAhg^{-1}$)	Structure	Method	Ref
Li-Si/C	1019	821 after 30 cycles at 160 mAg^{-1}	-	Li-Si alloys in carbon matrix	In-situ electrochemical	[20]
Si@graphitic	1479	1065 after 200 cycles at 280 mAg^{-1}	1042 at 23 Ag^{-1}	Nanosilicon-coated graphene granule	In-situ solid-state	[23]
Si@amorphous C	1291	650 after 100 cycles at 200 $mAhg^{-1}$	-	Nanosilicon amorphous carbon core-shell	In-situ carbothermal reduction	[27]
LRP-Si@C	2110	1633 after 70 cycles at 0.5 Ag^{-1}	580 at 8 Ag^{-1}	Lotus root-like porous	Magnesiothermic reduction and CVD	[28]
Microporous Si@C	1887	1210 after 40 cycles at 0.5 C	-	Nano core-shell	In-situ polymerization	[29]
Si@SiOx@C	1980	1450 after 100 cycles at 0.1 Ag^{-1}	1230 after 100 cycles (500 mAg^{-1})	Double-walled core-shell	Ball-milling and carbonization	[30]
Si/C	741.2	611.3 after 100 cycles at 0.3 Ag^{-1}	480.3 at 4 Ag^{-1}	Si/graphit@N-doped carbon core-shell	Spray-dying and carbonization	[31]

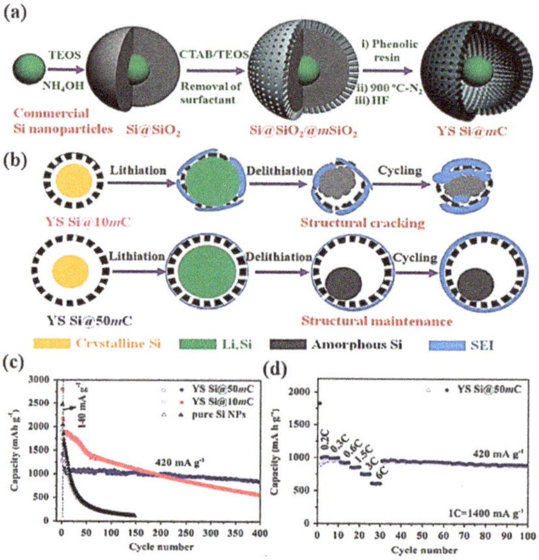

Figure 2. (**a**) Schematic illustration of the preparation of YS (yolk-shell) Si@mC; (**b**) schematic illustration of the lithiation–delithiation process of YS Si@10mC and YS Si@50mC; (**c**) cycling performance of pure Si NPs (nanoparticles), YS Si@10mC and YS Si@50mC; (**d**) rate performance of YS Si@50mC.

3.2. Embedded Composite

Because of the low silicon content of the embedded Si@C composite, its reversible capacity is relatively low; however, the high carbon content results in high stability [33]. The embedded composite refers to the embedding of silicon particles into the carbon matrix to form two particles, and the structure stability of the composite and the electrochemical activity of the electrode can be possibly improved depending on the conductive carbon medium. Wang et al. fabricated the embedded Si@C composite using a two-step CVD process, with which carbon nanotubes are grown, followed by the silica growth on carbon nanotubes. Through the physical force, silicon will embed into the carbon nanotube layered structure. The reversible capacity of the anode material was up to 2000 mAhg^{-1}, and after 40 cycles the capacity was basically unattenuated [34]. Lee et al [35] reported a Si-SiOx@C compound material with an embedded structure and a surface carbon layer, as illustrated in Figure 3. In the system, SiO_4^{4-} was transformed into a SiOx matrix by spray pyrolysis machine and citric acid was decomposed into a carbon layer. Therefore, the material presented high electrochemical performance (a reversible capacity of 1561 mAhg^{-1} at a rate of 0.06 C and, after 100 cycles, a capacity retention of 87.9% at a rate of 1 C). In the composite of Si–graphite, the cyclability is limited owing to the loose connections between flaked graphite particles and the poor interface adhesion among graphite and silicon. By introduction of another internal organic carbon source into the Si–graphite composite, the strong interfacial bonding can be formed and silicon can be evenly distributed into graphite. Datta et al. prepared a Si–graphite–PAN-C composite by the thermal decomposition of polyacrylonitrile (PAN)-based amorphous carbon (PAN-C), where the active Si phase was distributed in the graphite matrix. The introduced PAN acted as a diffusion barrier to inhibit the interfacial diffusion reaction between graphite and Si, so graphite retained its desired structure. This complete graphite was cut into ductile carbon matrix together with PAN-C to reduce the volume change of Si and suppress the irreversible loss of the anode. The composite exhibited a reversible capacity of 660 mAhg^{-1} with almost no decay up to 30 cycles at a constant current of 160 mAg^{-1}.

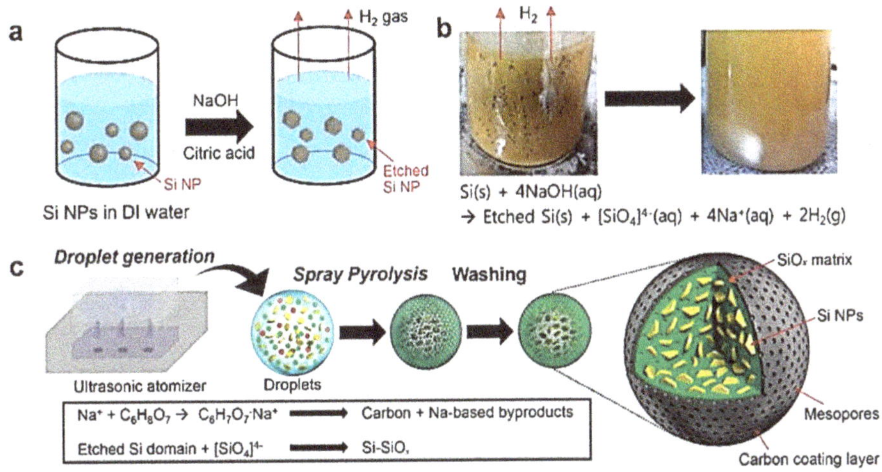

Figure 3. The synthesis illustration of the Si-SiOx@C compound material. (**a**) Schematic illustration of the preparation of the precursor solution. (**b**) Photographs of the precursor solution before and after the denoted reaction. (**c**) Graphical presentation for the synthesis of the Si−SiOx−C composite from the spray pyrolysis through a final washing step. Reprinted (adapted) with permission from [35], copyright 2017 American Chemical Society.

3.3. Doped Composite

The Silicon–amorphous carbon–graphite three elements Si@C composite system is mainly prepared by ball-milling and high-temperature pyrolysis. In this system, the chemical properties of the material can be improved by modifying the porous structure of silicon [36]. The existence of silicon contributes to the increase of capacity, graphite contributes to the improvement of the dispersion of silicon particles [37], and the amorphous carbon plays the role of the binder. In this composite, the type of carbon is changeable. The mesoporous carbon@ carbon nanotubes@ amorphous carbon materials are prepared using magnesium thermal reduction and CVD method. After 400 cycles, the reversible capacity remains 710 mAhg^{-1}. Wang et al. [38] reported Si@flake-graphite–amorphous-carbon by using ball-milling and spray drying. In this system, silicon and graphite were coupled by a PVP (polyvinyl pyrrolidone) binder, which was also used as a carbon source to form the coated layer. After high temperature treatment, a porous spherical shape was obtained. At a current density of 5 Ag^{-1}, the capacity of 200 mAhg^{-1} was maintained. After 300 cycles, the gravimetric capacity of up to 400 mAhg^{-1} was obtained. Wu et al. [39] used electrostatic spray deposition and heat treatment to synthesize a Si–graphene–porous carbon composite that had a layer-by-layer porous carbon framework to suppress the volume effect of silicon and a flexible graphene layer to facilitate electron transport and maintain integrity of the system. The material displayed a reversible capacity of 1020 mAhg^{-1} after 100 cycles at a current density of 200 mAg^{-1}. Agyeman et al. [40] synthesized the sandwich structure of Si@C-rGO by using stirring and vacuum filtration. Owing to the sandwich structure, with strong covalent and hydrogen bonding, the materials presented an excellent rate capacity of 767 mAhg^{-1} at a current density of 3 Ag^{-1}, high gravimetric capacity of 1001 mAhg^{-1} at 300 mAg^{-1}, and a great cyclic stability.

4. Conclusions and Prospect

The cycle stability and the reversible capacity retention rate are two important properties in the development of lithium ion batteries. It also restricts the commercialization of silicon carbon composites. By exploring the essential factors that affect its performance, the structure of composite materials is optimized to solve the problem of performance. In this paper, the recent research progress of in-situ synthesis of silicon carbon composites is described. The following suggestions are put forward for the future development of anode materials for lithium ion batteries: (1) Seeking a monomer with such kind of advantages, including the inorganic core and the organic shell of siliceous material; (2) optimizing the microstructure of composite material by a more effective preparation method; (3) combining the nano size with the mesoporous structure by the ternary system; (4) the combination of multiple forms achieves multiple performance capability.

In the system of in-situ electrochemical synthesis of Si@C anode materials, a new concept where the lithium-intercalated silicon alloy is uniformly dispersed in the carbon matrix materials has been proposed, which can make its capacity retention rate steady in a low voltage window without affecting the total energy density. By dispersing the lithium-intercalated alloys in different states into the matrix, and by preparing the anode materials with different contents of component, the electrochemical properties of the materials under different voltage windows can be investigated. Furthermore, it can also explain the lithium-intercalation mechanism of the anode materials in the process of charging–discharging. This method provides a new possibility for the preparation of anode materials and on the application and production with further development. In the system of in-situ solid-state synthesis, through the in-situ growth of graphene, the one-step solid-state synthesis of Si@C composites can be achieved. The composite takes full advantage of the excellent properties of graphene, which has high conductivity and can effectively regulate the charge–discharge cycle. This self-assembly method is simple, safe, and low-cost. It is helpful to synthesize graphene-based composites with different functions that can be applied to the lithium-ion batteries, supercapacitors, and fuel cells. In the system of in-situ carbothermal synthesis of Si@C anode materials, the raw materials are natural rice husk materials and the composites are prepared by a typical large-scale carbothermal reduction treatment. It is an environmentally friendly preparation method, which has the potential of large-scale production. Looking at the whole lithium-ion battery anode material industry, Si@C anode is still in the infancy stage, and its commercial application still needs further development. Nevertheless, in-situ synthesis provides a new direction for the preparation of Si@C anode materials. In the process of seeking high performance, the mechanism of internal electrochemical changes can also be studied, providing possibility for the improvement of the performance and industrialization.

Author Contributions: Conceptualization, W.T.; validation, H.Z.; investigation, Z.D.; writing—original draft preparation, Z.B.; writing—review and editing, H.T.

Funding: This research received no external funding.

Conflicts of Interest: The authors declare no conflict of interest.

References

1. Nishi, Y. Lithium ion secondary batteries; past 10 years and the future. *J. Power Sources* **2001**, *100*, 101–106. [CrossRef]
2. Goodenough, J.B.; Kyu-Sung, P. The Li-ion rechargeable battery: A perspective. *J. Am. Chem. Soc.* **2013**, *135*, 1167–1176. [CrossRef] [PubMed]
3. Boukamp, B.A.; Lesh, G.C.; Huggins, R.A. ChemInform Abstract: All-solid lithium electrodes with mixed-conductor matrix. *Chem. Informationsdienst* **1981**, *12*, 725–728. [CrossRef]
4. Chan, C.K.; Peng, H.; Liu, G.; McIlwrath, K.; Zhang, X.F.; Huggins, R.A.; Cui, Y. High-performance lithium battery anodes using silicon nanowires. *Nat. Nanotechnol.* **2008**, *3*, 31. [CrossRef] [PubMed]
5. Huggins, R.A. Lithium alloy negative electrodes. *J. Power Sources* **1999**, *81–82*, 13–19. [CrossRef]
6. Szczech, J.R.; Jin, S. Nanostructured silicon for high capacity lithium battery anodes. *Energy Environ. Sci.* **2010**, *4*, 56–72. [CrossRef]
7. Ng, S.H.; Wang, J.; Konstantinov, K.; Wexler, D.; Chew, S.Y.; Guo, Z.P.; Liu, H.K. Spray-pyrolyzed silicon/disordered carbon nanocomposites for lithium-ion battery anodes. *J. Power Sources* **2007**, *174*, 823–827. [CrossRef]
8. Ryu, J.H.; Kim, J.W.; Sung, Y.E.; Oh, S.M. Failure Modes of Silicon Powder Negative Electrode in Lithium Secondary Batteries. *Electrochem. Solid-State Lett.* **2004**, *7*, A306–A309. [CrossRef]

9. Polat, B.D.; Keles, O. Multi-layered Cu/Si nanorods and its use for lithium ion batteries. *J. Alloy. Compd.* **2015**, *622*, 418–425. [CrossRef]
10. Datta, M.K.; Kumta, P.N. Silicon and carbon based composite anodes for lithium ion batteries. *J. Power Sources* **2006**, *158*, 557–563. [CrossRef]
11. Lee, J.H.; Kim, W.J.; Kim, J.Y.; Lim, S.H.; Lee, S.M. Spherical silicon/graphite/carbon composites as anode material for lithium-ion batteries. *J. Power Sources* **2008**, *176*, 353–358. [CrossRef]
12. Yang, X.; Wen, Z.; Xu, X.; Lin, B.; Huang, S. Nanosized silicon-based composite derived by in situ mechanochemical reduction for lithium ion batteries. *J. Power Sources* **2007**, *164*, 880–884. [CrossRef]
13. Datta, M.K.; Kumta, P.N. In situ electrochemical synthesis of lithiated silicon-carbon based composites anode materials for lithium ion batteries. *J. Power Sources* **2009**, *194*, 1043–1052. [CrossRef]
14. Magasinski, A.; Dixon, P.; Hertzberg, B.; Kvit, A.; Ayala, J.; Yushin, G. High-performance lithium-ion anodes using a hierarchical bottom-up approach. *Nat. Mater.* **2010**, *9*, 353–358. [CrossRef]
15. Evanoff, K.; Magasinski, A.; Yang, J.; Yushin, G. Nanosilicon-Coated Graphene Granules as Anodes for Li-Ion Batteries. *Adv. Energy Mater.* **2011**, *1*, 495–498. [CrossRef]
16. Zhang, L.; Hao, W.; Wang, H.; Zhang, L.; Feng, X.; Zhang, Y.; Chen, W.; Pang, H.; Zheng, H. Porous graphene frame supported silicon@graphitic carbon via in situ solid-state synthesis for high-performance lithium-ion anodes. *J. Mater. Chem. A* **2013**, *1*, 7601–7611. [CrossRef]
17. Yang, X.-L.; Zhang, L.-L.; You, M.; Wen, Z.Y.; Wang, Q. Synthesis of Si/Sn binary lithium-storage host composite anode materials by in-situ mechanochemical reaction. *Chin. J. Inorg. Chem.* **2008**, *24*, 1320–1324.
18. Lee, H.Y.; Lee, S.M. Carbon-coated nano-Si dispersed oxides/graphite composites as anode material for lithium ion batteries. *Electrochem. Commun.* **2004**, *6*, 465–469. [CrossRef]
19. Zhang, P.C.; Yang, X.; Yu, D.-X.; Shi, C.-C.; Wen, Z.-Y. Synthesis of Silicon/Carbon Composite Anode Prepared by in-situ Carbothermal Reduction for Lithium Ion Batteries. *Chin. J. Inorg. Chem.* **2011**, *27*, 898–902.
20. Liu, D.; Zhang, W.; Lin, H.; Li, Y.; Lu, H.; Wang, Y. Hierarchical porous carbon based on the self-template structure of rice husk for high-performance supercapacitors. *RSC Adv.* **2015**, *5*, 19294–19300. [CrossRef]
21. Wang, L.; Schnepp, Z.; Titirici, M. Rice husk-derived carbon anodes for lithium ion batteries. *J. Mater. Chem. A* **2013**, *1*, 5269–5273. [CrossRef]
22. Xing, A.; Tian, S.; Tang, H.; Losic, D.; Bao, Z. Mesoporous silicon engineered by the reduction of biosilica from rice husk as a high-performance anode for lithium-ion batteries. *RSC Adv.* **2013**, *3*, 10145–10149. [CrossRef]
23. Jung, D.S.; Ryou, M.H.; Sung, Y.J.; Park, S.B.; Choi, J.W. Recycling rice husks for high-capacity lithium battery anodes. *Proc. Natl. Acad. Sci. USA* **2013**, *110*, 12229–12234. [CrossRef]
24. Ju, Y.; Tang, J.A.; Zhu, K.; Meng, Y.; Wang, C.; Chen, G.; Wei, Y.; Gao, Y. SiO_x/C composite from rice husks as an anode material for lithium-ion batteries. *Electrochim. Acta* **2016**, *191*, 411–416. [CrossRef]
25. Holzapfel, M.; Buqa, H.; Scheifele, W.; Novák, P.; Petrat, F.M. A new type of nano-sized silicon/carbon composite electrode for reversible lithium insertion. *Chem. Commun.* **2005**, *12*, 1566–1568. [CrossRef] [PubMed]
26. Holzapfel, M.; Buqa, H.; Krumeich, F.; Novák, P.; Petrat, F.M.; Veit, C. Chemical vapor deposited silicon/graphite compound material as negative electrode for lithium-ion batteries. *Electrochem. Solid-State Lett.* **2005**, *8*, A516–A520. [CrossRef]
27. Feng, X.; Yang, J.; Lu, Q.; Wang, J.; Nuli, Y. Facile approach to SiO_x/Si/C composite anode material from bulk SiO for lithium ion batteries. *Phys. Chem. Chem. Phys.* **2013**, *15*, 14420–14426. [CrossRef]
28. Lavoie, N.; Courtel, F.M.; Malenfant, P.R.L. Graphene-Based Composite Anodes for Lithium-Ion Batteries. *Nanostruct. Sci. Technol.* **2012**, 117–162.
29. Gao, P.; Fu, J.; Yang, J.; Lv, R.; Wang, J.; Nuli, Y.; Tang, X. Microporous carbon coated silicon core/shell nanocomposite via in situ polymerization for advanced Li-ion battery anode material. *Phys. Chem. Chem. Phys.* **2009**, *11*, 11101–11105. [CrossRef]
30. Wang, D.; Gao, M.; Pan, H.; Wang, J.; Liu, Y. High performance amorphous-Si@SiO_x/C composite anode materials for Li-ion batteries derived from ball-milling and in situ carbonization. *J. Power Sources* **2014**, *256*, 190–199. [CrossRef]
31. Zhou, R.; Guo, H.; Yang, Y.; Wang, Z.; Li, X.; Zhou, Y. N-doped carbon layer derived from polydopamine to improve the electrochemical performance of spray-dried Si/graphite composite anode material for lithium ion batteries. *J. Alloy. Compd.* **2016**, *689*, 130–137. [CrossRef]

32. Yang, J.; Wang, Y.X.; Chou, S.L.; Zhang, R.; Xu, Y.; Fan, J.; Zhang, W.-X.; Liu, H.K.; Zhao, D.; Dou, S.X. Yolk-shell silicon-mesoporous carbon anode with compact solid electrolyte interphase film for superior lithium-ion batteries. *Nano Energy* **2015**, *18*, 133–142. [CrossRef]
33. Yin, Y.X.; Xin, S.; Wan, L.J.; Li, C.J.; Guo, Y.G. Electrospray Synthesis of Silicon/Carbon Nanoporous Microspheres as Improved Anode Materials for Lithium-Ion Batteries. *J. Phys. Chem. C* **2011**, *115*, 14148–14154. [CrossRef]
34. Wei, W.; Kumta, P.N. Nanostructured hybrid silicon/carbon nanotube heterostructures: Reversible high-capacity lithium-ion anodes. *ACS Nano* **2010**, *4*, 2233–2241.
35. Lee, S.J.; Kim, H.J.; Hwang, T.H.; Choi, S.; Park, S.H.; Deniz, E.; Jung, D.S.; Choi, J.W. Delicate Structural Control of Si-SiO_x-C Composite via High Speed Spray Pyrolysis for Li-Ion Battery Anodes. *Nano Lett.* **2017**, *17*, 1870–1876. [CrossRef] [PubMed]
36. Zheng, Y.; Yang, J.; Wang, J.; NuLi, Y. Nano-porous Si/C composites for anode material of lithium-ion batteries. *Electrochim. Acta* **2007**, *52*, 5863–5867. [CrossRef]
37. Zuo, P.; Yin, G.; Ma, Y. Electrochemical stability of silicon/carbon composite anode for lithium ion batteries. *Electrochim. Acta* **2007**, *52*, 4878–4883. [CrossRef]
38. Wang, H.; Xie, J.; Zhang, S.; Cao, G.; Zhao, X. Scalable preparation of silicon@graphite/carbon microspheres as high-performance lithium-ion battery anode materials. *RSC Adv.* **2016**, *6*, 69882–69888. [CrossRef]
39. Wu, J.; Qin, X.; Zhang, H.; He, Y.B.; Li, B.; Ke, L.; Lv, W.; Du, H.; Yang, Q.-H.; Kang, F. Multilayered silicon embedded porous carbon/graphene hybrid film as a high performance anode. *Carbon* **2015**, *84*, 434–443. [CrossRef]
40. Agyeman, D.A.; Song, K.; Lee, G.H.; Park, M.; Kang, Y.M. Carbon-Coated Si Nanoparticles Anchored between Reduced Graphene Oxides as an Extremely Reversible Anode Material for High Energy-Density Li-Ion Battery. *Adv. Energy Mater.* **2016**, *6*, 1600904–1600913. [CrossRef]

© 2019 by the authors. Licensee MDPI, Basel, Switzerland. This article is an open access article distributed under the terms and conditions of the Creative Commons Attribution (CC BY) license (http://creativecommons.org/licenses/by/4.0/).

MDPI
St. Alban-Anlage 66
4052 Basel
Switzerland
Tel. +41 61 683 77 34
Fax +41 61 302 89 18
www.mdpi.com

Nanomaterials Editorial Office
E-mail: nanomaterials@mdpi.com
www.mdpi.com/journal/nanomaterials

www.ingramcontent.com/pod-product-compliance
Lightning Source LLC
Chambersburg PA
CBHW041951260926
51850CB00019B/2619
* 9 7 8 3 0 3 9 2 8 8 1 9 9 *